Scalar Field Cosmology

Series on the Foundations of Natural Science and Technology

ISSN: 2010-1961

Series Editors: C. Politis (*University of Patras, Greece*)
W. Schommers (*Forschungszentrum Karlsruhe, Germany*)
E. Meletis (*University of Texas at Arlington, USA*)

*For further details, please visit: http://www.worldscientific.com/series/sfnst

(Continued at end of book)

Series on the Foundations of Natural Science and Technology — Vol. 13

Scalar Field Cosmology

Sergei Chervon

Ulyanovsk State Pedagogical University, Russia

Igor Fomin

Bauman Moscow State Technical University, Russia

Valerian Yurov

Immanuel Kant Baltic Federal University, Russia

Artyom Yurov

Immanuel Kant Baltic Federal University, Russia

World Scientific

EW JERSEY · LONDON · SINGAPORE · BEIJING · SHANGHAI · HONG KONG · TAIPEI · CHENNAI · TOKYO

Published by

World Scientific Publishing Co. Pte. Ltd.

5 Toh Tuck Link, Singapore 596224

USA office: 27 Warren Street, Suite 401-402, Hackensack, NJ 07601

UK office: 57 Shelton Street, Covent Garden, London WC2H 9HE

British Library Cataloguing-in-Publication Data

A catalogue record for this book is available from the British Library.

Series on the Foundations of Natural Science and Technology — Vol. 13
SCALAR FIELD COSMOLOGY

Copyright © 2019 by World Scientific Publishing Co. Pte. Ltd.

ISBN 978-981-120-507-1

For any available supplementary material, please visit
https://www.worldscientific.com/worldscibooks/10.1142/11405#t=suppl

Desk Editor: Nur Syarfeena Binte Mohd Fauzi

Typeset by Stallion Press
Email: enquiries@stallionpress.com

Is dedicated to the bright memory of the wonderful scientists, our colleagues and elder friends Vitaly Melnikov and Pedro González-Díaz.

Preface

Cosmology has come a long way since its inception in the works of Albert Einstein. Although it started almost as a mathematical aside, a curious theoretical extrapolation of then new ideas of General Relativity Theory (GRT), it quickly caught the interest and imagination of a small but devoted circle of (mostly European) theoretical physicists, such as Alexander Friedmann (Russia), Georges Lemaître (Belgium) and Willem DeSitter (Netherlands). Despite working in relative obscurity, they have nevertheless managed the daunting task of finding some exact analytic solutions to the GRT equations and for the first time have demonstrated — to the utmost chagrin of Einstein — that under the normal physical conditions, the isotropic and homogeneous universe produces no stationary solutions. In other words, the universe must be evolving — expanding or contracting. However, as astounding as it could be, this visionary prediction at first gained almost no attention from the scientific community. There were two reasons for it: first, the works of Friedmann and Lemaître were published in relatively obscure European journals, whereas all the most powerful telescopes (and most wealthy observatories) were at the time situated in the United States — thus the papers were guaranteed to remain virtually unknown to the majority of the astronomers; second, those astronomers who did read it failed to see any evidence to support the "ridiculous" idea of a dynamically evolving universe. But everything changed in 1927, when the famous astronomer, Edwin Hubble, working with a 100-inch telescope in the Mt. Wilson observatory, made a startling discovery: the spectrum of all but the closest galaxies appeared to possess a characteristic red shift, directly proportional to the distances of these galaxies from ours. The resulting paper, entitled "A Relation Between Distance and Radial Velocity Among Extra-Galactic Nebulae", published by Hubble in 1929, in which he summarized this discovery in the form of the now-famous Hubble law, took physicists by storm. It was the undeniable, unequivocal proof that the observable universe undergoes the process of expansion. The dynamics of this expansion could only be explained by the solutions, derived by Friedmann and Lemaître, and being independently rediscovered by the American physicist Howard Robertson (his ideas would later be systematized and elaborated on by the British mathematician Arthur Geoffrey Walker, henceforth producing what amounts to one of the longest title in the history of science: the Friedmann-Lemaître-Robertson-Walker equations). Thus, cosmology produced the first properly verifiable prediction, that was triumphantly successful in explaining the observable universe.

The second prediction of the Friedmann, Lemaître, Robertson and Walker theory (FLRW) concerned the deep past of the universe, requiring that the universe must have an actual beginning — an initial singularity. The verification of this prediction proved to be rather tricky; it took decades for science to develop to a point where the physicists could with confidence predict the behaviour of extremely high energy plasma similar to the one that should have existed in the early, compact universe. First courageous foray into the matter was made in the 1948s by the theoretical physicists George Gamow and Ralph Alpher, who proposed that the light elements, such as hydrogen, helium and lithium, might have been produced in the early universe from an over-heated neutron gas via the mechanism akin to nuclear fusion. In retrospect, there were three remarkable things about their paper. First, the always joking Gamow could not resist to add a third name — that of Hans Bethe from Cornell University — to the list of the authors, reasoning that the paper dedicated to the creation of the matter as we know it should bear the names of people that began as Alpha, Betha and Gamma. Second, the detailed calculations done on the enormous computer belonging to the U.S. Bureau of Standards have shown that the model fails to produce anything heavier than helium. Third, and most importantly, the model predicted that the temperature of the universe must be about five degrees higher than 0 K. However, the failure of the theory to predict the relative abundance of the heavy elements quickly led the theory to obscurity. That is, until in 1965 the group of Princeton physicists (Robert Dicke, Jim Peebles, Peter G. Roll and David Todd Wilkinson) tried to answer the question: supposing that our observable expanding cosmos is merely a latest phase in a repeating sequence of the expansion/contraction of the universe, would there be any traces of the previous phases? To answer this question they had to study the intricacies of the dynamics of the early dense and super-hot universe — a sort of a stop-gap that separates the different phases of the universe. In the process they came to the same prediction the Gamow team had made almost 20 years prior — that there should be a residue primordial radiation, a sort of a 4–10 K temperature background, constant in every point of the universe. That same year two Soviet astronomers, Igor Novikov and Andrei Doroshkevich, have happened to stumble upon the aforementioned old paper by Gamow, Alpher and Bethe, and upon closer inspection have concluded that, although the model was wrong, its prediction about the background temperature were not. Furthermore, they have determined the increased background temperature to be observable by a sufficiently sensitive radio antenna, such as the microwave horn antenna, located in Holmdel, New Jersey, U.S. and owned by Bell Laboratories. While all that was happening, the Bell Lab's microwave horn antenna was occupied by a duo of physicists Arno Penzias and Robert Wilson, who were getting utterly exacerbated in trying to find the cause for an unexplained constant undirected low-frequency radio noise. After months of unsuccessful attempts to locate the source of the noise they were ready to give up until their colleague, Bernard Burke of the Carnegie Institution, have recommended Penzias to contact Dicke for a possible solution to the mystery of the noise. The result was a joint paper by Penzias and Wilson team and Dicke's group published in the Astrophysical Journal Letters in 1965, that not only proved the correctness of decades-old FLRW model up to the first few minutes of existence of the observable universe, but it also introduced a completely new instrument for studying the properties of the universe — the cosmic microwave background radiation (CMB).

However, once put to use, the CMB measurements soon led to a very unnerving conclusion. The extreme homogeneity of the CMB, originally taken as an ultimate proof of the correctness of FLRW model, was proven to actually be too extreme. The apparently casually disconnected regions of space had the CMB almost identical to each other. This implied that the early universe must have been practically almost entirely homogeneous and isotropic just to ensure that the causally disconnected regions evolve to the observable degrees of inhomogeneity. This strange incidence received a name "horizon problem" and, together with a number of other observational conundrums (most notably the flatness problem, and the problem of the large scale distribution of matter) led the physicists to a sinking realisation that something was seriously amiss with the accepted cosmological model. The only logical conclusion that could help to unravel the mystery was that at some early stage of its evolution the universe has had to expand exponentially fast. This was the only way to solve the problem of horizon (the currently disjoint regions of space actually were in contact with each other before the super-fast expansion and subsequent billions of years of standard FLRW dynamics have permanently separated them from each other), as well as the others. But what kind of field can possibly produce such monstrous dynamics? The answer was provided by the cosmologist Alan Guth, who in 1979 had identified the possible culprit as a scalar field, most probably associated with the supermassive Higgs boson. His idea was later elaborated and generalized by Andrei Linde, Alexei Starobinsky, Andreas Albrecht, Paul Steinhardt, and in its simplest form goes like this. Suppose the potential energy of interaction of the Higgs bosons (or other supermassive bosons with similar behaviour) has one of two possible shapes. If the temperature of the universe is too high (i.e. it exceeds the energy of grand unification $\Lambda_{GUT} \approx 10^{14}$ GeV, which will be the case in a very early universe), the potential has one global minimum, corresponding to a true vacuum state. But when the temperature reaches Λ_{GUT}, a second minimum might arise which, after the temperature sufficiently drops, becomes the new global minimum. The key point here lies in the realization that the potential between the two minima will at first be very shallow, hence ensuring that the scalar field will roll from the old true vacuum to the new one very slowly. Surprisingly, it is this slow-rolling condition that appears to be sufficient to produce the required exponential expansion of the universe, called the inflation, and it is the inflation that solved all the aforementioned problems tormenting cosmologists for decades. And this is why the delayed discovery of the Higgs boson in 2015 practically cemented the legacy of Guth and others and made this monograph possible and inevitable.

However, even then was not the end of the story. In the late 90s — early 00s two new breakthroughs once again threatened to revolutionize the contemporary cosmology. One was experimental. Another was theoretical. The experimental breakthrough came in 1998 with a discovery by two independent projects (the Supernova Cosmology Project and the High-Z Supernova Project) that about five billion years ago the universe began a new phase of the accelerated expansion. This discovery has completely undermined the previous ideas about the dominant types of matter in the universe. After the 1998 papers, cosmologists realized that the visible matter constitutes only 5 per cent of total matter count. Another 27 per cent appears to be a variation of a standard baryon matter that does not interact electromagnetically and thus cannot be directly observed — hence, the name "dark matter". And all the rest received the name of dark energy, and became a cause of much speculations as to its nature. Apparently, everything hinged on but one parameter — the barotropic correlation coefficient w between the density ρ and the pressure p of the dark matter. The accelerated expansion of the universe meant that $w < 0$, but just how far from zero was it? If $w = -1$ we end up with a familiar face — the vacuum energy, which ironically manifests itself in the FLRW equations as a term, specifically added by Einstein to produce a completely stationary universe. But if $w < -1$, then the dynamics of the universe becomes much more unusual. Such hypothetical fields has been called phantom fields, and we will talk lots more about the phantom cosmologies they induce in the later part of our monograph.

The second groundbreaking discovery was theoretical in nature and it stemmed from the attempts to connect two hitherto disjoint fundamental theories: the formalism of the quantum field theory and the geometric principles of the GRT. The first attempt at construction of such a theory — literally the theory of everything — was the theory of supergravity (SUGRA) in late 1970s, which was proven by Ed Witten to be consistent in a 11-dimensional space-time. Later the internal difficulties in the SUGRA implementation led physicists to reject it and replace with a 10-dimensional string theory (although it is now believed that the unified version of all self-consistent string theories — the M-theory — should actually be 11-dimensional, just as Witten has originally proposed). The additional dimensions, complementary to our observable 3 space + 1 time dimensions, might be invisible because they are microscopic in length (the Kaluza-Klein dimensions), but there appear to be another, much more interesting possibility: the brane model. According to it, while most of the additional dimensions are indeed microscopic, one of them is not; our 4-dimensional universe then behaves as a "brane" — a hypersurface embedded into a 5-dimensional bulk space; every physical field except gravitation is locked on the brane, so the only way the brane model might manifest itself (and therefore be observable) would be via cosmological dynamics (since it is governed primarily by the gravity). The first successful cosmological brane models were constructed in 1998 by Merab Gogberashvili and in 1999 by Lisa Randall and Raman Sundrum. We will look closer at the latter one in the last part of this book, as it most closely mimics the familiar FLRW model and thus serves as a wonderful arena for the generalization of the methods and approaches introduced in the earlier parts of the monograph.

With that said, let us now briefly discuss the structure of the book.

Part I "A canonical scalar field in Cosmology" essentially serves as an introduction to the subject. It was written with graduate students in mind, but we believe the material to also be useful for those researchers who might require some specific historic references, including the ones that are commonly (and rather unfairly) neglected, such as the fact that the representation of the scalar field dynamics in terms of the Hubble parameter H as a function of the scalar field was derived first by G. Ivanov in 1981 and then independently by D. Salopec and J. Bond in 1990.

In Chapter 1, Part I, we construct the generating function for scalar cosmology equations in the form of Ivanov-Salopec-Bond (ISB) representation and demonstrate the ways it can be used to reproduce exact solutions for various choices of the potential. The list of such potentials includes exponential, power-law, inverse-power, trigonometric and Higgs types of potentials. The new approach for solving ISB equation suggested by A. Muslimov in 1990 which is based on intermediate hyperbolic function is discussed as well.

The detailed description of the methods for the exact solutions construction is represented in the Chapter 2, Part I. We describe the method of fine tuning of the potential, originally introduced by G. Ellis and M. Madsen, and then discuss its applications. In particular, we demonstrate its usage by obtaining the inflationary solutions for the power-law, the exponential, the power-law-exponential and for the hyperbolic expansion of the Universe. Notably, in the case of the hyperbolic expansion, some of the solutions are completely new and original. On the next step, we consider the method of fine tuning of the potential in a case of a conformally flat spacetime. We also describe the method of the solution's construction from a given scalar field's evolution, which method was first proposed by J. Barrow in 1994. The method of obtaining the dual-generated solution, suggested by V. Zhuravlev and S. Chervon in 1998, will be described as well. The main feature of this method lies in gaining the second independent solution for the same potential without solving the dynamic equation. Finally, we describe the new exact hyperbolic solution that were produced by A. Chaadaev and S. Chervon in 2013 using the parametric representation for ISB equation with a cosmological constant.

Part II "Advanced methods of exact solutions' construction" is devoted to the approximation methods for construction of the analytic solutions and to the advanced methods of gaining the exact solutions. The slow-roll approximation we discussed above, and which above all provides an opportunity to relate the theoretical predictions with an observational data is presented in the following forms: the potential and the Hubble slow-roll representations.

The advanced methods of the exact solutions' construction are widely presented in this Part. The discussion of the method of "generating functions", introduced therein, is broken into five subsections, each corresponding to a particular class of functions, and being introduced in a chronological perspective. Special attention is devoted to a Schrödinger type equation, originally derived in this context by V. Zhuravlev, S. Chervon, V. Shchigolev in 1998. The results of investigations are represented in the table where comparison between the Schrödinger's picture and a standard cosmology was done. The relationship between the standard cosmology and the Ermakov-Pinney and Riccati equations is considered and the exact solutions are represented as well. The superpotential method in a standard cosmology with a scalar field, first suggested by S. Chervon and V. Zhuravlev in 2000, is represented along with the detailed examples of some of its exact solutions. Finally, the method of determination of differences between the exact and approximated solutions is presented.

Part III "Cosmological perturbations" includes the details of derivation of inflationary parameters and their classification as exact inflationary and slow-roll approximated parameters. Then the metric for tensor and scalar perturbations is presented in conformal time and the gauge invariant equations for linear perturbations on comoving hypersurfaces of constant energy density are wrote as well. Special attention and detailed derivation are devoted to quantum origin of the perturbations. Another issue concerns power spectra of scalar and tensor metric perturbation. It is shown how practically calculate spectral indexes, their runnings and tensor-to-scalar ratio on the crossing of Hubble radius. As the examples of method's application, calculations of cosmological parameters for power-law inflation, de Sitter solutions, generalized exponential and exponential-power-law inflation are performed. Analogy calculations are performed for conformally flat spacetime. Finally post-inflationary evolution of cosmological perturbations is analised.

Part IV, entitled "Friedmann vs Abel equations: A connection", discusses a very powerful method of obtaining the exact solutions for the scalar cosmology equations by transforming them to the Abel type differential equation. In particular, it is shown that the general solution of the Einstein-Friedman cosmological equations for the universe filled with a scalar field

φ of a known potential $V(\varphi)$ is explicitly connected to a general solution of a corresponding Abel equation of a first kind. This relationship can be formalised in terms of a strict mathematical theorem, the statement and the proof of which are both provided in the Part four. The theorem in question enables us to produce all cosmological parameters (including the scale factor and the scalar field) from a known scalar field potential and the solution of a corresponding Abel equation. The versatility of this technique is demonstrated on a number of models with physically meaningful potentials, including the models with a spontaneously broken symmetry. Furthermore, this formalism appears to be singularly useful for the tasks involving the study of the cosmological inflation. One of the examples provided therein is the analysis of the non-integrable quadratic model $V = m^2\varphi^2/2$, and the proof that this model not only naturally begets the inflation, but it also ends it in a natural fashion, without any need for fine tuning of the parameters of the model.

Part V "The Phantom Fields" is dedicated to the problem of the phantom fields in cosmology. The phantom fields belong to the domain of the most unusual and divisive concepts in contemporary cosmology. Taken in the framework of the quintessence models, the "phantom energy" can be described as a special sort of a scalar field with a negative kinetic term. Since such a field cannot exist in the classical field theory, we end up with a most unusual conclusion: the "phantom energy" should be an entirely quantum mechanical phenomenon, having no adequate solely classical description. Both the revolutionary and revelatory nature of this conclusion is obvious: should the future observations indeed confirm the existence of the phantom component (possessing the negative adiabatic parameter), this would be a first observable example of the quantum phenomena exerting the influence not even on the macro-, but on the mega-level. Furthermore, since the phantom energy violates the dominant-energy condition, it leads to very atypical cosmological dynamics, during which the energy density of the phantom component actually grew during the expansion of the universe. Although such a behaviour might seem to be strongly at odds with the laws of thermodynamics, the careful examinations of the problem demonstrated that the phantom energy should be characterized by a negative absolute temperature, thus, in an admittedly counterintuitive way, preserving both the first and second laws. For example, the entropy of the phantom component indeed increases as the universe expands!

On the other hand, the growth of the energy density of a "phantom field" during the expansion phase of the universe also means that the quantum gravity effects, so far deemed important only in the vicinity of the initial and/or final singularities $a \to 0$, can actually play a dominant role at the later stages of the cosmic evolution, and not merely on the planck lengths, but on the observable, macroscopic ones as well. Furthermore, the advent of the quantum gravity phenomena on the mega-level should lead to a number of very peculiar results, such as a possibility of causality violations on the macroscopic scale! And, in addition to all that, the phantom models possess a number of surprising properties that signify an interesting (albeit so far somewhat speculative) relationship between elementary particles physics and cosmology. The discussion of all these (and a few other, equally surprising) properties and predictions of the phantom cosmologies substitutes the core of the second half of the fifth Part of this monograph.

Part VI, "Branes", is devoted to the generalization of the methods of exact solution construction, originally developed for Friedman cosmology to the brane world cosmology. The very idea of the branes and, more broadly, of the macroscopic additional dimensions, has proven to be extremely powerful and influential in contemporary cosmology. The effectiveness of the approach is due to the fact that the adoption of its few underlying assumptions (that, in turn, are direct consequences of the string theory models) opens up a very simple and elegant way to solve the problem of the hierarchies. The key here lies in the concept of a three-dimensional brane embedded in the external $D + 4$-dimensional bulk space. In order to obtain the four-dimensional (time plus three spacial dimensions) action on the brane, one has to integrate the D-dimensional action with respect to the bulk space coordinates. As a result, in the simplest case the effective four-dimensional gravity constant G_4 arises as a result of division of its $D+4$-dimensional counterpart G_{D+4} over the corresponding D-dimensional volume. This, combined with appropriately large number D of the external macroscopic dimensions, can theoretically lead to the "abnormally" small effective quantity G_4, perfectly fitting the observations.

This approach can also help in resolving other cosmological problems; in particular, the problem of smallness of the cosmological constant λ. For this end, let us assume that the spatial volume of the bulk space is finite and has a characteristic size d. A common approach would be to impose an orbifold geometry on a bulk space, coupled with the Israel jump conditions imposed on the brane itself. The ensuing studies conclusively demonstrate that the effective (observable) value of the four-dimensional cosmological constant on the brane will be equal to its $D + 4$-dimensional counterpart, multiplied by an additional factor, proportional to e^{-d}. Hence, if the characteristic size of the bulk space is sufficiently large, the observable cosmological constant should become sufficiently small, in the process offering us a very elegant answer to the fundamental question of why the observable value of λ appears to be so scandalously small.

The detailed analysis of the problem has also provided two additional facts. First of all, the geometric configurations of the aforementioned type appears to be generally unstable. On the other hand, they can theoretically be stabilized via introduction of the additional scalar field to the bulk space. Subsequently constructed models, however, had one serious flaw: the solutions of their field equations were, generally speaking, singular both in the bulk space and on the brane. This provided an additional impetus to the problem of construction of such a mathematical algorithm that not only produces the exact solutions for the brane models, but also ensures that these solutions remain regular, stable and yield an exponentially suppressed value of the cosmological constant on the brane. Such a formalism can indeed be constructed, and the last section of Part VI of this monograph is dedicated to its derivation and application.

The authors

Contents

Part I

A canonical scalar field in cosmology

Chapter 1

Inflationary models with a canonical scalar field

1.1 Early Inflation and the implemented scalar field

The idea of *inflationary* expansion of the *universe's* evolution during very early times, once the *universe* emerged from the quantum gravity (Planck) era, has been proposed in the beginning of the 1980's and is becoming more popular as a necessary stage of the standard Big Bang theory model. The first works by Starobinsky (1980) [1], Guth (1981) [2], Linde (1982) [3], Albrecht and Steinhardt (1982) [4] include the physical mechanism based on quantum corrections and phase transitions during the very early stage of the *universe*. Exponential (de Sitter) expansion is the feature of *inflationary models* which helped to solve the long standing problems of the standard Big Bang theory model: the horizon, flatness, homogeneity and isotropy and some other problems. The chaotic *inflation* scenario proposed by Linde (1983) [5] differs from other previous versions since it is not based on the theory of high-temperature phase transition in the very early *universe*, but contains the locally homogeneous *scalar field* which is slowly rolling down to the minimum of the *scalar field potential*. After that proposal, many investigations were made on the *inflationary universe* connected with a self-interacting *scalar field* as the source of gravitation in the Friedmann world. Let us briefly mention some interesting works concerning the study of a *scalar field* in *inflationary cosmology*.

Homogeneous isotropic *cosmological models* with a massive *scalar field* have been studied in the works [6, 7]. It was shown that *inflationary* stages are a fairly general property of most solutions in the considered model. The general conditions for *inflation* were investigated in the work [8]. It was found that under the lower limit for the amplitude of a scalar field, the *universe* naturally enters into and exits out of an *inflationary* phase.

What is important is that such behavior takes place under a large variety of scalar potentials which are polynomial, logarithmic or exponential. It was also stated that a *scalar field* is essential for *inflation* [8]; it is unlikely that a vector or other non-scalar field will lead to *inflation*. The difference between scalar potentials in particle physics and these in *cosmology* has been stressed in the work [9]. The author wrote: "... we do not really know which theory of particle physics best describes the very early universe. One should therefore keep an open mind as to the form of $V(\phi)$." Halliwell chose the exponential potential and showed that it lead to the solution with power-law inflation and that this solution is an attractor. Detailed investigations of power-law inflation have been carried out in the work [10]. The authors found the constraints on the model coming from the requirement of solving the horizon, flatness, reheating and perturbation-spectrum problems. It was stated also that these constraints can be suitably satisfied. An exact power-law *inflationary* solution possessing an exponential *potential* was given in the work [11].

Let us mention also the investigation carried out by Ivanov (1981) [12] where he found the *exact solutions* for the nonlinear *scalar field* in *cosmology*. The solutions he obtained included polynomial, trigonometric and exponential potentials. The method he used for searching for *exact solutions* afterwards was called the *Hamilton-Jacobi-like approach*.

From the observational point of view, most results which can be related to observational data have been obtained from the so-called *slow-roll approximation* of the cosmological dynamical equations [3], [4]. Detailed investigations of various physical phenomena from particle physics and GUT theories for the period until the 1990's can be studied from the reviews [13–15]. Our attention will be concentrated on *exact solutions* of *inflationary* models, the study of which started about ten years later, after *inflationary cosmology* had been proposed.

Thus we are going to present a brief review of construction of *exact solutions* in the *inflationary universe*, i.e. the solutions of self-consistent Einstein and *scalar field* equations in Friedmann cosmology. The direct connection between *scalar field cosmology* and *cosmology* based on the *perfect fluid* stress-energy tensor needs to be mentioned. This connection is always valid except in the case of dust matter. Therefore we included the case of *exact solutions* for *perfect fluid* as the source of gravitation.

The construction of *exact solutions* in *inflationary cosmology* started with the work by Muslimov (1990) [16]. The results presented in this article will be discussed in Sec. 1.3. Here we would like to mention that the very method and many interesting *exact solutions* presented in [12] have been reproduced and generalized in [16]. New methods and new sets of *exact solutions* have been developed in the work [16] as well.

Barrow [17] found a simple way to solve exactly the cosmological dynamic equations in terms of a pressure-density relationship. In this way he obtained the known power-law and de Sitter forms of *inflation* and new classes of behavior in which the expansion *scale factor* increases as the exponent of some power of the cosmic time coordinate. The double-exponential law solution was obtained as well.

The work by Ellis and Madsen (1991) [18] was the first where "the inverse problem" was considered in the framework of *cosmology*. Usually one suggests that we know from HEP, the scalar *potential* in the very early *universe*, and our task is to find the *scale factor* and the *scalar field* as functions on time. However Ellis and Madsen (1991) [18] suggested starting from the given *scale factor* ! Indeed, it is clear that the *scale factor* may be found from observational data. Then we may take into account this fact to find the *potential* and *scalar field* from the cosmological equations. This work was done and examples of *exact solutions* have been presented for the pure *scalar field* (without taking into account radiation which is also considered there). Further this approach was developed in the works [19, 20]. A more detailed analysis of the so-called "*fine tuning potential method* " will be presented in Sec. 2.3.

1.2 Basic equations of scalar cosmology

We consider the model of a self-gravitating *scalar field* ϕ with the *potential* of self-interaction $V(\phi)$. The action of such a model is

$$S = \int d^4x \sqrt{-g} \left(\frac{R + \Lambda}{2\kappa} - \frac{1}{2}\phi_{,\mu}\phi_{,\nu}g^{\mu\nu} - V(\phi) \right), \qquad (1.1)$$

where R is the *curvature scalar*, ϕ the *scalar field*, $\phi_\mu = \partial_\mu \phi$ the short representation of the partial derivative $\partial\phi/\partial x^\mu$, κ is *Einstein's gravitational constant*, and Λ is the *cosmological constant*, which will mainly be included in the scalar field potential $V(\phi)$ as the constant part of it.

In the standard way one can obtain the *energy-momentum tensor* (EMT)

$$T_{\mu\nu}^{(sf)} = \phi_{,\mu}\phi_{,\nu} - g_{\mu\nu}\left(\frac{1}{2}\phi_{,\rho}\phi^{,\rho} + V(\phi) \right), \qquad (1.2)$$

and the *Einstein equation*

$$G_{\mu\nu} \equiv R_{\mu\nu} - \frac{1}{2}g_{\mu\nu}R = \kappa T_{\mu\nu}^{(sf)}, \tag{1.3}$$

may be represented through the trace of the EMT in the form

$$R_{\mu\nu} = \kappa \left(-T_{\mu\nu} + \frac{1}{2}g_{\mu\nu}T \right) = \phi_{,\mu}\phi_{,\nu} + g_{\mu\nu}V(\phi). \tag{1.4}$$

Varying the action (1.1) with the scalar field ϕ, we obtain the dynamic equation of the *scalar field*

$$-\nabla_\mu \nabla^\mu \phi + V'(\phi) = 0, \quad V' \equiv \frac{dV}{d\phi}. \tag{1.5}$$

We consider the homogeneous and isotropic *universe* as the spacetime with the *Friedmann-Robertson-Walker (FRW) metric*

$$ds^2 = -dt^2 + a^2(t)\left(\frac{dr^2}{1 - \epsilon r^2} + r^2\left(d\theta^2 + \sin^2\theta d\varphi^2 \right) \right), \tag{1.6}$$

where $\epsilon = 0$, $\epsilon = 1$, $\epsilon = -1$ for the spatially-flat, closed and open *universe*, respectively.

The Einstein equation (1.3) and the equation of the *scalar field* dynamics (1.5) in the FRW metric (1.6) lead to the system of equations

$$\frac{\ddot{a}}{a} + \frac{2\dot{a}^2}{a^2} + \frac{2\epsilon}{a^2} = \kappa V(\phi), \tag{1.7}$$

$$-\frac{3\ddot{a}}{a} = \kappa \left(\dot{\phi}^2 - V(\phi) \right), \tag{1.8}$$

$$\ddot{\phi} + 3\frac{\dot{a}}{a}\dot{\phi} + V'(\phi) = 0. \tag{1.9}$$

Equations (1.7) and (1.8) can, in an equivalent way, be replaced by their sum and the linear combination $3 \times (1.7) + (1.8)$. Including the *Hubble parameter* $H = \dot{a}/a$, $\dot{a} = da/dt$, the system (1.7)–(1.9) can be rewritten in the form

$$H^2 + \frac{\epsilon}{a^2} = \frac{\kappa}{3}\left(\frac{1}{2}\dot{\phi}^2 + V(\phi) \right), \tag{1.10}$$

$$\dot{H} - \frac{\epsilon}{a^2} = -\kappa\frac{1}{2}\dot{\phi}^2, \tag{1.11}$$

$$\ddot{\phi} + 3H\dot{\phi} + V'(\phi) = 0. \tag{1.12}$$

We will refer to the system (1.10)–(1.12) as *the Scalar Cosmology Equations (SCEs)*.

The representation above, Eqs. (1.10)–(1.12) has some advantage in being able to derive any one (1.10)–(1.12) from the other two, and differential consequences of each case.

Another representation of the SCEs was first proposed by G. Ivanov [12]. Suggesting the dependence of the *Hubble parameter* H on the *scalar field* ϕ, the transformation of the equations (1.10)–(1.12) for the spatially-flat *universe* ($\epsilon = 0$) to the form, which was called later the Hamilton-Jacobi-like form, was made. Equation (1.11) is transformed to

$$H' = -\kappa \frac{1}{2}\dot{\phi}. \tag{1.13}$$

Squaring the above equation and making the substitution $\dot{\phi}^2/2$, expressed in term of H'^2, and substituting into (1.10), one can obtain

$$\frac{2}{3\kappa}\left[\frac{dH}{d\phi}\right]^2 - H^2 = -\frac{\kappa}{3}V(\phi). \tag{1.14}$$

It is worthwhile to note that this procedure and the very equation (1.14) have been obtained by G. Ivanov in 1981 [12]. Unfortunately this result was published in limited editions (in Russian) and it was not familiar outside the USSR. Fortunately, in 1990, in the work of A. Muslimov [16], the Ivanov result was reproduced (Muslimov referenced the Ivanov article) and some of the solutions were generalized as well. In the same year, within the detailed investigation of long-wavelength metric fluctuations in *inflationary* models, D. Salopek and J. Bond [21] obtained the "separated *Hamilton-Jacobi equation* that also governs the semiclassical phase of the wave functional". The obtained equation contains a couple of *scalar fields* and definitely can be applied to a single *scalar field*. Therefore we suggest that Eq. (1.14) in the cosmological context should be called *Ivanov-Salopek-Bond (ISB) equation*.

1.3 Generating function for solving the Ivanov-Salopek-Bond equation

The structure of Eq. (1.14) prompts us to find the form of the *potential* which can give the *exact solution*. Indeed, let the *potential* $V(\phi)$ be of the form

$$V(\phi) = -\frac{2}{3\kappa}F'^2 + (F(\phi) + F_*)^2, \tag{1.15}$$

where $F = F(\phi)$ is a C^1 function on ϕ and $F_* = const$. Following the terminology of Sec. 4.1, we will call the function $F(\phi)$ *the generating function*.

Comparing Eq. (1.14) with (1.15), it is easy to find the solution of (1.14)

$$H(\phi) = \sqrt{\frac{\kappa}{3}}(F(\phi) + F_*). \tag{1.16}$$

Thus, one can directly from (1.14) obtain the *potential* if the *Hubble parameter* is given. And vice versa, if one sets the *potential* in the form (1.15), then the solution of (1.14) will be defined by Eq. (1.16).

1.3.1   *Potential in polynomial form*

As an example let us choose the *generating function* $F(\phi)$ as the finite series on degrees of the field ϕ

$$F(\phi) = \sum_{k=0}^{p} \lambda_k \phi^k + F_*. \tag{1.17}$$

Under this circumstance, the potential $V(\phi)$ takes the following form

$$V(\phi) = -\frac{2}{3\kappa}\left[\sum_{k=0}^{p-1}(k+1)\lambda_{k+1}\phi^k\right]^2 + \left[\sum_{k=0}^{p}\lambda_k \phi^k + F_*\right]^2. \tag{1.18}$$

Let us consider the simple case when $F_* = 0$, $k = 0$, $p = 1$. Then the *potential* becomes

$$V(\phi) = -\frac{2}{3\kappa}\lambda_1^2 + \lambda_0^2 + 2\lambda_0\lambda_1\phi + \lambda_1^2\phi^2. \tag{1.19}$$

The *generating function* $F(\phi)$ and the *Hubble parameter* are

$$F(\phi) = \lambda_0 + \lambda_1\phi, \quad H(\phi) = \sqrt{\frac{\kappa}{3}}(\lambda_0 + \lambda_1\phi). \tag{1.20}$$

If we additionally set $\lambda_0 = 0$, then we obtain the solution for the massive *scalar field* as in [12], with $\lambda_1^2 = m^2/2$ (for the sake of simplicity we also set $c = \hbar = 1$). Thus the *potential* takes the form

$$V(\phi) = \frac{m^2\phi^2}{2} - \frac{m^2}{3\kappa}. \tag{1.21}$$

Solving Eq. (1.13) one can obtain the evolution of the *scalar field*

$$\phi(t) = -m\sqrt{\frac{2}{3\kappa}}t + \phi_s = -m\sqrt{\frac{2}{3\kappa}}(t - t_*), \quad \phi_s = m\sqrt{\frac{2}{3\kappa}}t_*. \tag{1.22}$$

Index "s" ("*singularity*") here is related to the values at the initial time $t = 0$, i.e. for *singularity* in accordance with Big Bang theory. The *Hubble parameter*

$$H = m\sqrt{\frac{\kappa}{6}}\phi, \tag{1.23}$$

having the dependence ϕ on time (1.22), gives a possibility to perform integration and obtain the dependence of the *scale factor* on time

$$a = a_s \exp\left(-\frac{m^2}{6}t^2 + m\sqrt{\frac{\kappa}{6}}\phi_s t\right). \tag{1.24}$$

Thus we obtained the *exact solution* for the potential (1.21), which represented by the dependence ϕ on time (1.22) and the *scale factor* on time (1.24).

It is interesting to note that the same solution and its application for calculation of *e-folds number* and scalar spectral parameter, were found and developed later by Wang [22].

When $\lambda_0 \neq 0$ the solution for the *scale factor* will differ by the factor a_s in front of the exponent

$$a = a_s \exp\left(-\frac{m^2}{6}t^2 + m\sqrt{\frac{\kappa}{6}}\phi_s t + \lambda_0\sqrt{\frac{\kappa}{3}}\right)$$

$$= \tilde{a}_s \exp\left(-\frac{m^2}{6}t^2 + m\sqrt{\frac{\kappa}{6}}\phi_s t\right). \tag{1.25}$$

Here $\tilde{a}_s = a_s \exp\left(\lambda_0\sqrt{\kappa/3}\right)$. The *potential* $V(\phi)$ then takes the form

$$V(\phi) = -\frac{2}{3\kappa}\lambda_1^2 + \lambda_0^2 + 2\lambda_0\lambda_1\phi + \lambda_1^2\phi^2 = \left(\frac{m\phi}{\sqrt{2}} + \lambda_0\right)^2 - \frac{m^2}{3\kappa}. \tag{1.26}$$

Let us note that the linear transformation of the field without changing of the mass

$$\tilde{\phi} = \frac{\phi}{\sqrt{2}} + \frac{\lambda_0}{m} \tag{1.27}$$

leads to the potential (1.21) for the field $\tilde{\phi}$.

Let us consider the case when $k = 1, 2$. The *generating function* $F(\phi)$ takes the form

$$F(\phi) = \lambda_1\phi + \lambda_2\phi^2. \tag{1.28}$$

The *potential* $V(\phi)$ is

$$V(\phi) = -\frac{2}{3\kappa}(\lambda_1 + 2\lambda_2\phi)^2 + (\lambda_1\phi + \lambda_2\phi^2)^2. \qquad (1.29)$$

We can make a simplification by considering $\lambda_1 = 0$. The *potential* then takes the *Higgs* form

$$V(\phi) = -\frac{2}{3\kappa}(2\lambda_2\phi)^2 + (\lambda_2\phi^2)^2. \qquad (1.30)$$

Using the relation (1.16), we may find H and H' expressed through ϕ

$$H = \sqrt{\frac{\kappa}{3}}\lambda_2\phi^2, \quad H' = 2\sqrt{\frac{\kappa}{3}}\phi. \qquad (1.31)$$

Equation (1.13) takes the form

$$2\sqrt{\frac{\kappa}{3}}\phi = -\frac{\kappa}{2}\dot{\phi}. \qquad (1.32)$$

Performing the integration we find the dependence of ϕ on time

$$\phi = \exp\left[-\frac{4}{\sqrt{3\kappa}}\lambda_2(t - t_*)\right]. \qquad (1.33)$$

Substituting this result into (1.31) and performing the integration, we will find the *scale factor* $a(t)$

$$a(t) = a_s \exp\left[-\frac{\lambda_2\kappa}{8}\exp\left(-\frac{8\lambda_2}{\sqrt{3\kappa}}(t - t_*)\right)\right]. \qquad (1.34)$$

This is the double-exponental law solution [12], [17].

   To make a comparison with Ivanov's results (with subscript "I" in his notation) [12], let us display the relations between the parameters of the model

$$\mu_I = -\frac{16}{3\kappa}\lambda_2^2, \quad \lambda_I = -4\lambda_2^2, \quad 3\kappa\mu_I = 4\lambda_I.$$

The case when $\lambda_1 \neq 0$ leads to the *scalar field*

$$\phi = \frac{1}{2\lambda_2}\exp\left[-\frac{4}{\sqrt{3\kappa}}\lambda_2(t - t_*) - \frac{\lambda_1}{4\lambda_2}\right]. \qquad (1.35)$$

The *scale factor* then takes the following form

$$a(t) = a_s \exp\left[-\sqrt{\frac{\kappa}{3}}\frac{\lambda_1^2}{4\lambda_2}t - \frac{\sqrt{3\kappa}}{8\lambda_2}\exp\left(-\frac{8\lambda_2}{\sqrt{3\kappa}}(t - t_*)\right)\right]. \qquad (1.36)$$

It is useful to note the role of the addition of the constant F_* to the function $F(\phi)$. In our presentation for $H(\phi)$ (1.16), we can extract the constant part of the *Hubble parameter*

$$H(\phi) = \sqrt{\frac{\kappa}{3}}(F + F_*) = \tilde{H}(\phi) + H_*. \tag{1.37}$$

The presence of the constant H_* will be exhibited as the additional factor for $a(t)$

$$a(t) = e^{H_* t} \exp\left(\int H dt\right). \tag{1.38}$$

In the considered examples above (1.25) and (1.36), such factors can be extracted explicitly.

1.3.2 *Trigonometric potential*

The solution for the potential which leads to the *Sine-Gordon type equation* was obtained in [12]. Such a setting used the special choice of the additional parameter. Let us consider this point in detail.

Corresponding *generating function* $F(\phi)$, we choose as

$$F(\phi) = A\sin(\lambda\phi), \quad A, \ \lambda - const. \tag{1.39}$$

Then the *potential* is

$$V(\phi) = -\frac{2A^2\lambda^2}{3\kappa}\cos^2(\lambda\phi) + A^2\sin^2(\lambda\phi). \tag{1.40}$$

To obtain the potential suggested in [12], it is enough to choose the parameter λ in the following way: $\lambda^2 = \frac{3\kappa}{2}$. Such a choice leads to the *potential*

$$V(\phi) = -A^2\cos\left(\sqrt{6\kappa}\phi\right). \tag{1.41}$$

Equating the parameter $A^2 = \mu$, we find the correspondence of the potential function with that presented in [12].

The *Hubble parameter* in terms of the *scalar field* can be defined from (1.16)

$$H(\phi) = A\sqrt{\frac{\kappa}{3}}\sin(\lambda\phi). \tag{1.42}$$

Integrating the equation (1.13) it is necessary to consider the integral

$$\int \frac{dx}{\cos x}$$

which has various functional representations

$$\int \frac{dx}{\cos x} = \ln\left|\tan\left(\frac{\pi}{4} + \frac{x}{2}\right)\right| + c_1 = \frac{1}{2}\ln\frac{1 + \sin x}{1 - \sin x} + c_2, \quad c_1, c_2 - const.$$
(1.43)

In [12] the first representation was shown. We take the second representation in which the *scalar field* is defined from the relation

$$\sin(\lambda\phi) = \tanh\left(A\lambda\sqrt{2}(t - t_*)\right).$$
(1.44)

Applying this result to (1.42) and performing the integration over time t, we obtain the *scale factor*

$$a(t) = a_s\left[\cosh(A\lambda\sqrt{2}(t - t_*))\right]^{1/3} = a_s\left[\cosh(A\sqrt{3\kappa}(t - t_*))\right]^{1/3}.$$
(1.45)

1.3.3 *Exponential potential*

The exponential potential in [12] is represented as

$$V(\varphi) = \alpha\exp(\beta\varphi), \quad \text{where } \alpha, \ \beta - const.$$
(1.46)

If we set

$$F(\phi) = A\exp(\mu\phi),$$
(1.47)

then the *potential* takes the form

$$V(\phi) = A^2\left(1 - \frac{2\mu^2}{3\kappa}\right)\exp(2\mu\phi).$$
(1.48)

Comparing this result to the original *potential* (1.46), we can find the relations

$$\alpha = A^2\left(1 - \frac{2\mu^2}{3\kappa}\right), \quad \beta = 2\mu, \quad A = \sqrt{\frac{\alpha}{1 - \frac{\beta^2}{6\kappa}}}.$$
(1.49)

In accordance with the general procedure explained in the beginning of Sect. 1.3, one can obtain

$$H(\phi) = \sqrt{\frac{\kappa}{3}}A\exp(\mu\phi).$$
(1.50)

Then the dependence of the *scalar field* on time t has a logarithmic character

$$\phi(t) = -\frac{1}{\mu}\ln\left(\frac{2A\mu^2}{\sqrt{3\kappa}}t\right) + \phi_s. \tag{1.51}$$

The *scale factor* is evaluated via a power law

$$a(t) = a_s(t - t_*)^{\kappa/2\mu^2}. \tag{1.52}$$

An addition of the constant F_* to $F(\phi)$ leads to the generalization of the solution (1.52)

$$a(t) = a_s e^{H_* t}(t - t_*)^{\kappa/2\mu^2}, \quad H_* = \sqrt{\frac{\kappa}{3}}F_*. \tag{1.53}$$

This is the exponential power law solution. Then the *potential* acquires the additional terms

$$V(\phi) = A^2\left(1 - \frac{2\mu^2}{3\kappa}\right)\exp(2\mu\phi) + 2AF_* e^{\mu\phi} + F_*^2. \tag{1.54}$$

Muslimov (1990) [16] found the generalization of Ivanov's solution for the exponential *potential*. Let us represent this, which contains both solutions.

If we take the *generating function* $F(\phi)$ in the form (1.47) with the *potential* (1.54) then

$$H(\phi) = \sqrt{\frac{\kappa}{3}}\left(Ae^{\mu\phi} + F_*\right). \tag{1.55}$$

We can find that

$$H' = \frac{\kappa}{3}A\mu e^{\mu\phi}. \tag{1.56}$$

Integrating (1.13), we obtain

$$e^{\mu\phi} = \left(\frac{2A\mu^2}{\sqrt{3\kappa}}t + v_*\right)^{-1}, \tag{1.57}$$

where v_* is a constant of integration. Finally we find

$$\phi(t) = -\frac{1}{\mu}\ln\left(\frac{2A\mu^2}{\sqrt{3\kappa}}t + v_*\right). \tag{1.58}$$

To obtain Ivanov's solution [12] we set $v_* = 0$ and take into account the relations (1.49).

To obtain Muslimov's solution [16] we set $v_* = 1$ and take into account the relations below

$$F_* = 0, \quad \Lambda = A^2\left(1 - \frac{2\mu^2}{3\kappa}\right), \quad A = 2/\mu, \quad B = A/\sqrt{3}.$$

The solution (1.58) without restrictions on the parameter v_* gives some generalization.

1.3.4   *The solution with an inverse potential*

The *potential* in [16] was presented in the following way

$$V(\phi) = m^2 \phi^{-\beta} \left(1 - \frac{1}{6}\beta^2 \phi^{-2} \right), \quad \beta > 0. \tag{1.59}$$

It is not difficult to check that the same *potential* can be obtained from the *generating function*

$$F(\phi) = m\phi^{-\beta/2}. \tag{1.60}$$

Therefore it is clear that the solution can be obtained by the general scheme. The *Hubble parameter* is

$$H(\phi) = \sqrt{\frac{\kappa}{3}} m\phi^{-\beta/2} + H_*. \tag{1.61}$$

As we know the influence of H_* on the result, we may take $H_* = 0$ for the sake of simplicity.

Integrating (1.13) we can find the dependence of the *scalar field* on time

$$\phi(t) = [K_1(t - t_*)]^{2/(\beta+4)} + \phi_*, \quad K_1 = \sqrt{\frac{\kappa}{3}} \left(\frac{\beta+4}{2m\beta} \right). \tag{1.62}$$

This result leads to the time dependence of the *Hubble parameter*

$$H(t) = \sqrt{\frac{\kappa}{3}} m \left[K_1(t - t_*) \right]^{-\beta/(\beta+4)}. \tag{1.63}$$

Then the *scale factor* is

$$a = a_s m \sqrt{\frac{\kappa}{3}} \left(\frac{\beta+4}{4} \right) K_1^{-\beta/(\beta+4)} \exp\left((t - t_*)^{4/(\beta+4)} \right). \tag{1.64}$$

The solution of such a type can be confronted with both a very early and late time *universe*. Similar solutions have been obtained in [17].

1.3.5   *The solution with an intermediate (hyperbolic) function*

In the range of the results described above, Muslimov [16] suggested a new original approach for solving the *scalar field cosmology* equation. To simplify calculations, let us, following [16], introduce a new variable

$$x = \sqrt{\frac{3\kappa}{2}}\phi, \tag{1.65}$$

and the potential function

$$f^2 = \frac{\kappa}{3}|V(\phi)|. \tag{1.66}$$

For this notation, the *Ivanov-Salopek-Bond (ISB) equation* (1.14) reduces to

$$(H'_x)^2 - H^2 = \mp f^2 \tag{1.67}$$

The upper case corresponds to the positive sign of the *potential*. It is interesting to mention that an equation of this type was studied by Mitrinovitch in 1937 [23].

Let us search for a solution in the form

$$H(x) = f(x)\cosh\left(\coth^{-1} y(x)\right), \quad y > 1. \tag{1.68}$$

The other choice is for the lower sign

$$H(x) = f(x)\sinh\left(\tanh^{-1} y(x)\right), \quad y < 1. \tag{1.69}$$

Here we included inverse hyperbolical tangents instead of inverse hyperbolical cotangents as in Muslimov's work with the aim of avoiding plus-minus signs in the final equation. Using the formulae above, for transition to the function $y(x)$, we obtain

$$[f'\cosh u + f\sinh u\, u']^2 - [f(x)\cosh u]^2 = -f^2. \tag{1.70}$$

Here, for the sake of briefness, we introduce the function u in the following way

$$u(y) = \coth^{-1} y(x). \tag{1.71}$$

We can shift the second term on the left hand side of (1.70) to the right hand side and, using the property of the hyperbolic function, take the root on the left and right hand side of the equation (considering all values as positive). As the result, we obtain the equation

$$f'\cosh u + f\sinh u\, u' = f\sinh u. \tag{1.72}$$

The transition to the function $y(x)$ is performed by inverse substitution of (1.71) and by disclosing the derivative

$$u' = \frac{y'}{1 - y^2}.$$ (1.73)

Finally we arrive at the following relation

$$\frac{f'}{f} = \tanh(\coth^{-1} y)\left(1 - \frac{y'}{1 - y^2}\right).$$ (1.74)

After simple algebraic transformations we acquire the *Abel equation*

$$y' = \frac{f'}{f}y^3 - y^2 - \frac{f'}{f}y + 1.$$ (1.75)

Repeating the same procedure for the lower case, we arrive once again to Eq. (1.75).

Chapter 2

Methods of exact solutions' construction

2.1 Fine tuning of the potential

Analysing the system of cosmological equations we can ask the question: "What do we know from experiment?". The well-known answer is "The *potential* $V(\phi)$". We know this from high energy physics (HEP). A similar argument may be related to kinetic energy or an evolutionary law of the *scalar field*. Whilst it is correct when we are speaking about the very early universe, what we do know about the cosmological *potential*? First of all, if we are ready to accept the fact that radiation-dominated and matter-dominated stages existed in the history of the evolution of the *universe*, we may say that the *potential* has an exponential form (in the sense that we are describing these stages by a *scalar field* rather than by a *perfect fluid*). A similar argument may be related to the kinetic energy or an evolutionary law of the *scalar field*. Studying the evolution of the present *universe*, we include *dark energy* (DE) to explain the acceleration of the *universe*. This means that the cosmological *potential* may have other forms as a functional dependence on the *scalar field* ϕ.

Ellis and Madsen [18] first suggested assuming the *scale factor* of the *universe* as obtained from observations or by reasonable cosmological behavior: "The dynamics we consider here are governed by standard general relativity new families of solutions should be obtained by techniques introduced here. The potential we obtain ad hoc in the sense that they are derived from the desired behaviour of the *universe*, rather than from a field theory model. One aims to use the present analysis to determine the *potential* that gives the best behaviour in terms of its implications for *cosmology*".

In [18] the representation of the SCEs (in the absence of radiation, which was also taken into consideration in this work) was in the form

$$V(\phi) = \dot{H} + 3H^2 + \frac{2\epsilon}{a^2}, \tag{2.1}$$

$$K(\phi) \equiv \frac{1}{2}\dot{\phi}^2 = -\dot{H} + \frac{\epsilon}{a^2}, \tag{2.2}$$

in natural units in which $8\pi G_N = 1$.

The solutions have been constructed in [18] for the following *scale factors*:

- $a(t) = A \exp(\lambda t)$
- $a(t) = A \sinh(\lambda t)$
- $a(t) = A \cosh(\lambda t)$
- $a(t) = A t^m$

Let us explain the method in detail following the articles [24], [20] and reconstruct the solutions for the *scale factors* listed above.

For the sake of convenience let us reproduce the SCEs (1.7)–(1.9) of the model (1.1) in the FRW spaces (1.6)

$$\frac{\ddot{a}}{a} + \frac{2\dot{a}^2}{a^2} + \frac{2\epsilon}{a^2} = -\Lambda + V(\phi), \tag{2.3}$$

$$-\frac{3\ddot{a}}{a} = \Lambda + \left(\dot{\phi}^2 - V(\phi)\right), \tag{2.4}$$

$$\ddot{\phi} + 3\frac{\dot{a}}{a}\dot{\phi} + \frac{dV(\phi)}{d\phi} = 0. \tag{2.5}$$

Here, *Einstein's gravitational constant* is put equal to unity: $\kappa = 1$ and the cosmological constant Λ is extracted from the *potential* $V(\phi)$.

To obtain a new class of *exact solutions*, we will make use of the freedom in the choice of the form of the *potential*.

Considering the Eqs. (2.3)–(2.5), one can find that the last Eq. (2.5) is the differential consequences of (2.3) and (2.4). To prove this fact, one can differentiate (2.3) wrt t to obtain the following equation

$$\frac{\dddot{a}}{a} + 3\frac{\dot{a}\ddot{a}}{a^2} - \frac{4\dot{a}^3}{a^3} - \frac{4\dot{a}\epsilon}{a^3} - \frac{dV}{dt} = 0. \tag{2.6}$$

This equation can be rewritten as

$$\frac{\dddot{a}}{a} - 3\frac{\dot{a}\ddot{a}}{a^2} + \frac{2\dot{a}^3}{a^3} + \frac{2\dot{a}\epsilon}{a^3} - \quad (2.7)$$

$$-6\frac{\dot{a}}{a}\left(-\frac{\ddot{a}}{a} + \frac{\dot{a}^2}{a^2} + \frac{\epsilon}{a^2}\right) - \frac{dV}{dt} = 0.$$

Now one can use two consequences of the Einstein equations (2.3) and (2.4):

- the sum of *Einstein's equations* (2.3) and (2.4) firstly is

$$\dot{\phi}^2 = \frac{2}{a^2}\left(-a\ddot{a} + \dot{a}^2 + \epsilon\right), \quad (2.8)$$

- and, secondly, the time derivative from the relation (2.8) is

$$\ddot{\phi}\dot{\phi} = \partial_t\left(-\frac{\ddot{a}}{a} + \frac{\dot{a}^2}{a^2} + \frac{\epsilon}{a^2}\right). \quad (2.9)$$

Inserting the left hand side of (2.8) and (2.9) into (2.8) and using $dV/dt = (dV/d\phi)\dot{\phi}$, we can divide the equations (2.8) by $\dot{\phi} \neq 0$. As a result, one obtains the equation for the *scalar field* (2.5).

The following analysis will be used just for *Einstein's equations* (2.3) and (2.4), which can be reduced to the form where the functions $V(t) \equiv V(\phi(t))$ and $\phi(t)$ are expressed through the function $a(t)$ and their derivatives [19]:

$$V(t) = \Lambda + \frac{\ddot{a}}{a} + \frac{2\dot{a}^2}{a^2} + \frac{2\epsilon}{a^2}, \quad (2.10)$$

$$\phi(t) = \pm\sqrt{2}\int\left(\sqrt{-\frac{d^2\ln a}{dt^2} + \frac{\epsilon}{a^2}}\right)dt. \quad (2.11)$$

By giving the rate of the expansion as a functional dependence of the *scale factor* on time, $a = a(t)$, we can find the functions $\phi(t)$ and $V(t)$ which support a chosen type of evolution. It is obvious that the pair of functions (2.10) and (2.11) give the parametric dependence $V = V(\phi)$. In some cases, after the calculation of the right hand sides in (2.10), (2.11), it is possible to find the explicit dependence $V = V(\phi)$ by eliminating t.

2.2   The inflationary solutions

The *exact solutions* for the exponential and power law type of *inflation* in the framework of 'fine tuning of the potential' (FTP) method have been obtained in [19, 20, 25].

2.2.1   *Power law inflation*

To obtain power law *inflation* let us start from the assumption that

$$a(t) = At^m, \quad A = const. \tag{2.12}$$

where $m > 1$. The integral on the right hand side of (2.11) can be calculated explicitly. Therefore we find ($m \neq 1$)

$$V = \Lambda + \frac{m}{t^2}(3m - 1) + \frac{2\epsilon}{A^2 t^{2m}}, \tag{2.13}$$

$$\phi(t) = \pm \sqrt{\frac{1}{2} \frac{m}{(1-m)}} \left\{ 2\sqrt{1 + \alpha t^{-2m+2}} \right.$$

$$\left. + \ln \left(\frac{\sqrt{1 + \alpha t^{-2m+2}} + 1}{\sqrt{1 + \alpha t^{-2m+2}} - 1} \right) \right\} + \phi_0, \tag{2.14}$$

where $\alpha = \epsilon A^{-2}/m$.

In the case of the spatially-flat *universe* ($\epsilon = 0$) the solution for arbitrary m has the form

$$\phi = \pm\sqrt{2m} \ln t + \phi_0, \tag{2.15}$$

$$V(t) = \Lambda + \frac{m + 3m^2}{t^2}. \tag{2.16}$$

Eliminating t, we find an exponential dependence of V on ϕ

$$V(\phi) = \Lambda + (m + 3m^2)e^{-\sqrt{2\kappa m^{-1}}(\phi - \phi_0)}, \tag{2.17}$$

which is usually the definition of power law *inflation* [13]. In the case of the open and closed *universe* ($\epsilon \neq 0$) it is possible to find an explicit dependence V on ϕ only for some values of m. For example, if $m = 1$ (in this case the formulas (2.13) and (2.14) do not work),

$$V(\phi) = \Lambda + \exp\left(-\frac{2(\phi - \phi_0)}{\pm\sqrt{2}\sqrt{1 + \epsilon A^{-2}}}\right). \tag{2.18}$$

2.2.2 de Sitter exponential expansion

Let us present the set of solutions obtained by the *fine tuning potential method*. We start from the solutions found in the work [18]. We choose the evolution of the scale factor as

$$a(t) = A \exp(\lambda t), \quad A, \ \lambda - const. \tag{2.19}$$

As a result one can find

$$H(t) = \lambda, \quad \dot{H} = 0, \quad V(t) = 3\lambda^2 + \frac{\epsilon}{A^2} e^{-2\lambda t}. \tag{2.20}$$

The scalar field dynamics are given by

$$\phi(t) = \mp \frac{\sqrt{2\epsilon}}{A\lambda} e^{-\lambda t} + \phi_s, \quad \phi_s - const. \tag{2.21}$$

Here we may consider $\epsilon = -1$ as the solution corresponding to a *phantom scalar field.*

Expressing the time t as a function of the *scalar field* ϕ from (2.21) and making the substitution in (2.20), one can obtain the *potential V* as a function of ϕ as

$$V(\phi) = 3\lambda^2 + |\epsilon| \lambda^2 (\phi - \phi_s)^2. \tag{2.22}$$

Here we used the symbol $|\epsilon|$ to stress that the *potential* is not equal to a constant only for open and closed *universes.*

2.2.3 Exponential-power-law inflation

Let us generalize the previous two cases and consider the evolution of the *scale factor* in the form

$$a(t) = A \exp(\lambda t) t^m, \quad \text{where } A, \ m, \ \lambda - const. \tag{2.23}$$

The *Hubble parameter* and its first derivative are

$$H(t) = \lambda + \frac{m}{t}, \quad \dot{H} = -\frac{m}{t^2}. \tag{2.24}$$

The *scalar field* dynamics are described by

$$\phi(t) = \mp \sqrt{2m} \ln t + \phi_s, \quad \text{where } \phi_s = const., \tag{2.25}$$

$$V(t) = 3\lambda^2 + \frac{6\lambda m}{t} + \frac{m(3m - 1)}{t^2}. \tag{2.26}$$

Using the relation (2.26) between the *scalar field* and cosmic time, it is not difficult to find the dependence of V on ϕ:

$$V(\phi) = 3\lambda^2 + 6\lambda m \exp\left(\mp\frac{\phi - \phi_s}{\sqrt{2m}}\right)$$

$$+ m(3m - 1)\exp\left(\mp\frac{2(\phi - \phi_s)}{\sqrt{2m}}\right). \tag{2.27}$$

As we can see, this *potential* falls under the *generating function* method described above in Sec.1.3.

2.2.4 *Hyperbolic inflation*

The case when the *scale factor* $a(t)$ of the *universe* grows exponentially very fast has been analysed in [19, 26]. Let us mention the differences in Barrow's and the FTP approaches. To find new *exact solutions* with Barrow's method [26], one needs to take the *scalar field* as a function of time $\phi = \phi(t)$, and then determine the evolution of the *scale factor* $a(t)$ and the *potential* $V(\phi(t))$ from it. This approach was applied in the works [18, 27, 28] as well. With the FTP method [19, 20] the *exact solutions* have been obtained by taking, first, the *scale factor* as the function of time $a = a(t)$ and then determining the evolution of the *potential* $V = V(t)$, and the evolution of a *scalar field* $\phi = \phi(t)$. The dependence between V and ϕ is, in general, a parametric one. Thus we can combine both of the methods, if we put the metric obtained by Barrow's way as the seed solution in the FTP method.

As an example of a synthesis of both methods, let us consider one of the *exact solutions* obtained in [26]. Following the suggested method, let us consider the *De Sitter expansion* from a *singularity* using the *scale factor* in the 'sinh' form

$$a(t) = A\sinh^\alpha(\lambda t), \quad A, \ \alpha, \ \lambda - -const. \tag{2.28}$$

It is easy to find $H(t)$ and $\dot{H}$

$$H(t) = \alpha\lambda\coth(\lambda t), \quad \dot{H} = -\frac{\alpha\lambda^2}{\sinh^2(\lambda t)}. \tag{2.29}$$

Then, by integrating Eq. (2.2) one can find the solution for the *scalar field*

$$\phi - \phi_* = \pm\sqrt{2\alpha}\ln\left|\tanh\left(\frac{\lambda t}{2}\right)\right| = \pm\sqrt{2\alpha}\ln\left|\frac{e^{\lambda t} - 1}{e^{\lambda t} + 1}\right|. \tag{2.30}$$

The *history of the potential*, i.e., the dependence of V on the cosmic time t, may be presented in two forms

$$V(t) = \alpha\lambda^2(3\alpha - 1)\coth^2(\lambda t) + \alpha\lambda^2, \tag{2.31}$$

or

$$V(t) = \alpha\lambda^2(3\alpha - 1)\sinh^{-2}(\lambda t) + 3\alpha^2\lambda^2. \tag{2.32}$$

Expressing the time t from the solution of the *scalar field* (2.30), one can find the *potential* as a function of ϕ

$$V(\phi) = \alpha\lambda^2(3\alpha - 1)\sinh^2\left(\frac{\phi}{\pm\sqrt{2\alpha}}\right) + 3\alpha^2\lambda^2. \tag{2.33}$$

This result coincides with the solution obtained in [18] if we set $\alpha = 1$.

If we compare the 'sinh α' solution presented here with that obtained on the basis of the *scalar field* evolution in the work [26], then for the case $\alpha = 1$ we can use re-parametrization in the following way

$$a_0 = A, \quad A = \pm\sqrt{2\alpha}, \quad \lambda = \lambda/2, \quad \varphi = \phi. \tag{2.34}$$

This leads to the same solution.

The case of 'cosh α' evolution can be considered as *De Sitter expansion* without a *singularity* [18]. We choose the *scale factor* in the form

$$a(t) = A\cosh^\alpha(\lambda t), \quad \text{where } A, \ \alpha, \ \lambda - const. \tag{2.35}$$

The *Hubble parameter* and its derivative are

$$H = \alpha\lambda\tanh(\lambda t), \quad \dot{H} = \alpha\lambda^2\cosh^{-2}(\lambda t). \tag{2.36}$$

The equation for the *scalar field* definition

$$\frac{1}{2}\dot{\phi}^2 = -\dot{H}$$

takes the form

$$\dot{\phi}^2 = -2\alpha\lambda^2\cosh^{-2}(\lambda t). \tag{2.37}$$

The power α should be positive (in the opposite case we will have a contracting *universe*). Therefore, a given expansion in the case of the flat *universe* may be supported by a *phantom scalar field*. Therefore we will change the sign of the kinetic term and the equation for the *scalar field* definition transforms to

$$\dot{\phi}^2 = 2\alpha\lambda^2\cosh^{-2}(\lambda t). \tag{2.38}$$

Note, that in the work [18], a closed *universe* was considered the case of the 'cosh' evolution ($\alpha = 1$).

Taking square roots and integrating the resulting equation from (2.38) one can find the evolution for the *scalar field* as:

$$\phi = \pm 2\sqrt{2\alpha}\arctan\left(e^{\lambda t}\right).\tag{2.39}$$

The *potential* as a function of time has the following form

$$V(t) = \alpha\lambda^2(1 - 3\alpha)\cosh^{-2}(\lambda t) + 3\alpha^2\lambda^2.\tag{2.40}$$

Using (2.39), it is not difficult to express the *potential* as a function of ϕ

$$V(\phi) = \alpha\lambda^2(1 - 3\alpha)\sin^2\left(\frac{\phi}{\pm\sqrt{2\alpha}}\right) + 3\alpha^2\lambda^2.\tag{2.41}$$

2.3 Fine tuning method in a conformally flat space

In this section we present the *fine tuning potential method* for a conformally flat space-time which is often using for consideration of *cosmological perturbations* [29]. We may represent the FRW metric for a spatially flat model in a conformally flat form

$$ds^2 = a^2(\eta)[-d\eta^2 + (dx^1)^2 + (dx^2)^2 + (dx^3)^2].\tag{2.42}$$

The SCEs (1.7)–(1.9) in terms of *conformal time* ($d\eta = dt/a$) and conformal *Hubble parameter* $\mathcal{H} := (da/d\eta)(1/a)$ can be rewritten as follows:

$$3\mathcal{H}^2 = \frac{1}{2}\phi'^2 + a^2V(\phi),\tag{2.43}$$

$$\phi'' + 2\mathcal{H}\phi' + a^2\frac{dV(\phi)}{d\phi} = 0,\tag{2.44}$$

$$\mathcal{H}' - \mathcal{H}^2 = -\frac{1}{2}\phi'^2.\tag{2.45}$$

Here and below the prime denotes a derivative with respect to the *conformal time*.

Using the definition of $\mathcal{H}$, from equation (2.45), we can obtain

$$\frac{1}{2}\phi'^2 = -\frac{a''}{a} + 2\left(\frac{a'}{a}\right)^2.\tag{2.46}$$

Substituting (2.46) in Eq. (2.43), and using the definition of $\mathcal{H}$, after simple transformations, we can write the *potential* $V(\phi)$ as

$$V(\phi(\eta)) = \frac{a''}{a^3} + \frac{a'^2}{a^4}.\tag{2.47}$$

To simplify Eqs. (2.46)–(2.47), we will carry out a transformation for the conformal scale factor: $a(\eta) = \sqrt{-A(\eta)}$. Then the FRW metric takes the form

$$ds^2 = A(\eta)[d\eta^2 - (dx^1)^2 - (dx^2)^2 - (dx^3)^2]. \tag{2.48}$$

Substituting the relation $a(\eta) = \sqrt{-A(\eta)}$ into Eqs. (2.46) and (2.47), we obtain expressions for the *potential* and the *scalar field* via the conformal scale factor $A(\eta)$:

$$V(\phi(\eta)) = -\frac{A''}{2A^2}, \tag{2.49}$$

$$(\phi')^2 = -\frac{A''}{A} + \frac{3}{2}\left(\frac{A'}{A}\right)^2. \tag{2.50}$$

For ease of calculation, we consider the metric (2.48) with signature $(-,+,+,+)$, which means the replacement of $A(\eta)$ with $-A(\eta)$. In this case, $A(\eta) = a^2(\eta)$, the metric is

$$ds^2 = A(\eta)[-d\eta^2 + (dx^1)^2 + (dx^2)^2 + (dx^3)^2] \tag{2.51}$$

and Eqs. (2.49)–(2.50) are written in the following form

$$V(\phi(\eta)) = \frac{A''}{2A^2}, \tag{2.52}$$

$$(\phi')^2 = -\frac{A''}{A} + \frac{3}{2}\left(\frac{A'}{A}\right)^2, \tag{2.53}$$

$$\mathcal{H} = \frac{A'}{2A}. \tag{2.54}$$

The representation above allows us to obtain *exact solutions* of the SCE's by choosing $A(\eta)$.

2.3.1 *Conformal exponential expansion*

Now, we consider the *conformal factor* $A(\eta) = A_0 e^{\beta(\eta)}$. From Eqs. (2.52)–(2.54) we obtain *exact solutions* of the SCEs

$$V(\phi(\eta)) = \frac{e^{-\beta(\eta)}}{2A_0}[\beta''(\eta) + \beta'^2(\eta)], \tag{2.55}$$

$$\phi(\eta) = \pm \int \sqrt{\frac{\beta'^2(\eta)}{2} - \beta''(\eta)} \, d\eta + \phi_0, \tag{2.56}$$

$$\mathcal{H}(\eta) = \frac{\beta'(\eta)}{2}, \tag{2.57}$$

$$a(\eta) = \sqrt{A_0}\, e^{\beta(\eta)/2}. \tag{2.58}$$

Thus, setting the function $\beta = \beta(\eta)$, we can obtain *exact solutions* for specific *cosmological models*.

2.3.2   Conformal power-law expansion

Let us represent the *conformal factor* as $A(\eta) = A_0 e^{m \ln(\alpha\eta)} = A_0 (\alpha\eta)^m$, $\beta(\eta) = m \ln(\alpha\eta)$. Then, from the Eqs. (2.55)–(2.58) we obtain the solution

$$\phi(\eta) = \pm\sqrt{\frac{m(m+2)}{2}}\,\ln(\alpha\eta) + \phi_0, \tag{2.59}$$

$$\eta = \frac{1}{\alpha} e^{\pm\sqrt{\frac{2}{m(m+2)}}(\phi-\phi_0)}, \tag{2.60}$$

$$V(\phi) = \frac{m(m-2)}{4A_0}\, e^{\mp\sqrt{\frac{2(m+2)}{m}}(\phi-\phi_0)}, \tag{2.61}$$

$$a(\eta) = \sqrt{A_0}\,\alpha^{m/2}\eta^{m/2} = \tilde{a}_0 \eta^{m/2}. \tag{2.62}$$

2.3.3   Generalized conformal exponential expansion

Let us choose the *conformal factor* in the form $A(\eta) = A_0 e^{\beta(\eta)}$, where $\beta(\eta) = \frac{c_1}{2}\eta^2 + c_2\eta + c_3$. The solution in this case can be written as:

$$V(\phi(\eta)) = \frac{e^{-\beta}}{2A_0}[c_1 + (c_1\eta + c_2)^2], \tag{2.63}$$

$$\phi(\eta) = \pm\frac{\sqrt{2}}{c_1}\left[\frac{\tilde{\eta}}{2}\sqrt{\tilde{\eta}^2 - c_1} - \frac{c_1}{2}\ln|\tilde{\eta} + \sqrt{\tilde{\eta}^2 - c_1}|\right], \tag{2.64}$$

where $\tilde{\eta} = (c_1\eta + c_2)/\sqrt{2}$. If we define the scale factor as

$$a(\eta) = \sqrt{A_0}\, e^{\frac{c_1}{4}\eta^2 + \frac{c_2}{2}\eta + \frac{c_3}{2}}, \tag{2.65}$$

then, for the function $\beta(\eta) = \frac{b_1}{12}\eta^4 + b_3$, we obtain

$$V(\phi(\eta)) = \frac{e^{-\beta}}{2A_0}\left(b_1\eta^2 + \frac{b_1^2}{9}\eta^6\right), \tag{2.66}$$

$$\phi(\eta) = \pm\frac{3}{2\sqrt{2b_1}}\left[\frac{\tilde{\eta}}{2}\sqrt{\tilde{\eta}^2 - b_1} - \frac{b_1}{2}\ln|\tilde{\eta} + \sqrt{\tilde{\eta}^2 - b_1}|\right], \tag{2.67}$$

$$a(\eta) = \sqrt{A_0} e^{\frac{b_1}{24}\eta^4 + \frac{b_3}{2}}. \tag{2.68}$$

where $\tilde{\eta} = (b_1\eta^2)/3$.

2.4 Solution construction from a scalar field evolution

Let us follow the method suggested by Barrow in the work [26]. We choose a spatially-flat *universe* and Eqs. (1.7)–(1.8) as the basic ones. As it was shown in Sec. 1.2, the Eq. (1.9) can be derived from (1.7)–(1.8).

The construction of an *exact solution* using the law of *scalar field* evolution will be presented for one special case. Let us set the functional dependence of the *scalar field* on time t as in [26]

$$\phi(t) = A \ln[\tanh(\lambda t)]. \tag{2.69}$$

Then, from (1.11) with $\epsilon = 0$, we can calculate the derivative of the *Hubble parameter* $\dot{H}$:

$$\dot{H} = -\frac{1}{2}A^2\lambda^2 \sinh^{-2}(\lambda t)\cosh^{-2}(\lambda t) \tag{2.70}$$

and, integrating the last equation, one can find

$$H = A^2\lambda \coth(2\lambda t), \tag{2.71}$$

or, as a function of ϕ,

$$H(\phi) = \lambda A^2 \cosh(\phi/A). \tag{2.72}$$

Thus, the scale factor is

$$a(t) = a_s \left[\sinh(2\lambda t)\right]^{A^2/2}. \tag{2.73}$$

The potential can be obtained from substituting the RHS (1.11) into (1.10), and expressing $V(\phi)$ from the resulting formula. In the given case the *potential* is

$$V(\phi) = A^2\lambda^2 \left[(3A^2 - 2)\cosh^2(\phi/A) + 2\right]. \tag{2.74}$$

In the work [26] there were investigations of the following evolutionary laws of *scalar fields*

$$\phi(t) = A \ln[\tanh(\lambda t)], \tag{2.75}$$

$$\phi(t) = A \ln[\tan(\lambda t)], \tag{2.76}$$

$$\phi(t) = A/\sin(\lambda t). \tag{2.77}$$

It is interesting to note that in the work [30], Barrow mentioned a few evolutions of the *scalar field* in discussions on physical *potentials* and relations between the *slow-roll approximation* and *exact solutions*. The evolution of the *scalar field* is represented by the following dependencies

$$\phi(t) = A \ln t, \tag{2.78}$$

$$\phi(t) = A t^m, \tag{2.79}$$

$$\phi(t) = A(\ln t - B)^n. \tag{2.80}$$

Solutions obtained by the *fine tuning potential method* in [18] contain the following evolutions of the *scalar field*

$$\phi(t) = A e^{-\lambda t}, \tag{2.81}$$

$$\phi(t) = A \ln \left[\frac{e^{\lambda t} - 1}{e^{\lambda t} + 1} \right], \tag{2.82}$$

$$\phi(t) = A \arctan(e^{\lambda t}), \tag{2.83}$$

$$\phi(t) = A \ln t. \tag{2.84}$$

It is clear that (2.82) and (2.83) have the same evolution.

2.5 A dual-generated solution construction

We are going to analyse the possibility of generating *exact solutions* for a given history of the *potential*, i.e., when the *potential* is represented as a function of time: $V = V(t)$. Generally speaking, if we know the dependencies $V(t)$ and $\phi(t)$, we may follow the standard view when the *potential* is a function of the *scalar field* $V = V(\phi)$. To this end, one may use each function $V(t)$ and $\phi(t)$ as parametric representations of a particular function $V(\phi)$. In the work [24], the function $V(t)$ was called *the potential history* when they were considering the overall evolution of the *universe* instead of the *inflationary* phase only.

Let us transform the *potential* history equation

$$V(t) = \frac{\ddot{a}}{a} + 2\frac{\dot{a}^2}{a^2} + 2\frac{\epsilon}{a^2} + \Lambda, \tag{2.85}$$

using the new function $Z(t) = a^3(t)$. The resulting equation is

$$\ddot{Z} + 3 \left[\Lambda - V(t) \right] Z + 6\epsilon Z^{1/3} = 0. \tag{2.86}$$

In the case of a spatially-flat *universe* ($\epsilon = 0$), this equation acquires the form of an ordinary linear differential equation

$$\ddot{Z} + 3\left[\Lambda - V(t)\right] Z = 0. \tag{2.87}$$

This equation has the same form as the Schrödinger one for the motion of a quantum particle in one-dimensional space with a particle potential energy $U(t) = 3\kappa V(t)$ and self-energy $E = 3\Lambda$.

Because of the structure of Eq. (2.87), we can state that if we know one of the solutions of this equation for a fixed dependence $V(t)$, we can derive another solution without solving Eq. (2.87) once again. Therefore we may term this solution *the dual-generated* solution. Let us denote the solution by $Z_1(t)$ which was obtained, say by the FTP method. A second linearly independent solution can be denoted by $Z_2(t)$ for the same fixed function $V(t)$ or $U(t)$. Then we may use the relation between any two independent solutions of the Eq. (2.87) [31] in the form

$$Z_1\dot{Z}_2 - Z_2\dot{Z}_1 = w_*, \ w_* = const. \tag{2.88}$$

Here and below the letter with the subscript ($_*$) means a constant. From Eq. (2.88) one can obtain the solution Z_2

$$Z_2 = Z_1 \left(c_1 + w_* \int \frac{dt}{Z_1^2} \right), \tag{2.89}$$

where c_1 is an integration constant.

Let us (following the article [24]) consider the result of the application of the method described above. There the solution for a minimally varying *scalar field* was considered. The first solution for the *scale factor* is

$$a_1(t) = a_s(a_*t + b_*)^{1/a_*^2} \exp\left[c_*t\right]. \tag{2.90}$$

The scalar field evolution is defined by

$$\phi(t) = \sqrt{2/\kappa}\, a_*^{-1} \ln(a_*t + b_*) + \phi_* \tag{2.91}$$

and the *potential* is

$$V(\phi) = \kappa^{-1}\left(\Lambda + 3c_*^2 + \left(\frac{3}{a_*^2} - 1\right)\exp\left[-2\alpha(\phi - \phi_*)\right]\right.$$
$$\left. + 6\frac{c_*}{a_*}\exp\left[-\alpha(\phi - \phi_*)\right]\right), \tag{2.92}$$

where $\alpha = a_*\sqrt{\kappa/2}$.

A dual to this solution, $a_2(t)$, can be obtained from the formula

$$a_2(t) = a_1(t) \left(c_1 + a_s^{-6} w_* \int dt (a_* t + b_*)^{-6/a_*^2} \exp\left[-6 c_* t\right] \right)^{1/3}. \quad (2.93)$$

The simplification taken, for the purposes of easy integration in the article [24], was $c_* = 0$. After that the solution has the form

$$a_2^{(pl)}(t) = a_s (a_* t + b_*)^{1/a_*^2} \left(c_1 + a_s^{-6} w_* \frac{a_*^2}{a_*^2 - 6} (a_* t + b_*)^{1-6/a_*^2} \right)^{1/3}. \quad (2.94)$$

Simplifying the *scale factor* to

$$a_1(t) = a_s t^m, \quad (2.95)$$

we obtain the solution $a_2(t)$ in the form

$$a_2(t) = a_s t^m \left(c_1 + a_s^{-6} (1 - 6m)^{-1} t^{(1-6m)} \right)^{1/3}. \quad (2.96)$$

Setting $c_1 = 0$, the solution reads

$$a_2(t) = a_s^{-5} (1 - 6m)^{-1} t^{(1/3 - m)}. \quad (2.97)$$

Thus we can see that for the power law *scale factor* $a_1(t) = a_s t^m$, the dual solution will be represented also as a power law solution with the power $(1/3 - m)$. This means that for the *inflationary* solution with $m > 1$, the dual solution leads to a contracting *universe*.

2.6   Hyperbolic solution for $H(\phi)$

Let us turn back to the *Ivanov-Salopek-Bond equation* (1.14). After including the *cosmological constant* Λ, the ISB equation can be transformed to the following form

$$\alpha H'^2 - \beta H^2 + V(\phi) = \kappa^{-1} \Lambda, \quad (2.98)$$

where $\alpha = 2/\kappa^2$, $\beta = 3/\kappa$. Let us study the case when, according to weak energy condition, the potential energy is positive with its constant constituent $V_0 = \kappa^{-1}\Lambda$, i.e., $V(\phi) - V_0 > 0$. Introducing new variables

$$y = H', \quad x = H, \quad B^2(\phi) = \frac{V(\phi) - V_0}{\alpha}, \quad A^2(\phi) = \frac{V(\phi) - V_0}{\beta}, \quad (2.99)$$

Eq. (2.99) is reduced to a hyperbolic canonical equation with variable coefficients which depend on the scalar field ϕ:

$$\left(\frac{x(\phi)}{A(\phi)} \right)^2 - \left(\frac{y(\phi)}{B(\phi)} \right)^2 = 1. \quad (2.100)$$

Equation (2.100) has the standard representation in the parametric form

$$\begin{cases} x = A\cosh(\lambda\phi), \\ y = B\sinh(\lambda\phi). \end{cases} \tag{2.101}$$

This system is consistent only for a certain value of $V(\phi)$. Using the definition of x and y, we obtain the obvious relation $x' = y$. This relation leads to the restriction on the *potential*

$$V(\phi) = V_0 + C\left(\cosh(\lambda\phi)\right)^{2(\delta-1)}, \tag{2.102}$$

where $\delta = \lambda^{-1}\sqrt{3\kappa/2} = const$. In the case of violation of the weak energy condition, we have $V(\phi) - V_0 < 0$. The procedure described above gives the restriction on the potential in the form

$$V(\phi) = V_0 + C\left(\sinh(\lambda\phi)\right)^{2(\delta-1)}. \tag{2.103}$$

It is interesting to note that the obtained solution can be reduced, with the special choice of parameters [32], to the solution found in the work [33].

Our next task is to find a dependence of the *Hubble parameter* or the *scale factor* on the cosmic time t. To this end we will use the equations

$$H' = -\frac{\kappa}{2}\dot{\phi}, \tag{2.104}$$

where

$$H(\phi) = \sqrt{\frac{\kappa C}{3}}\left(\cosh(\lambda\phi)\right)^\delta.$$

Introducing the notation $\xi = \sinh(\lambda\phi)$, Eq. (2.104) may be transformed to the following integral

$$\int \left(1+\xi^2\right)^{-\frac{\delta}{2}}\frac{d\xi}{\xi} = -\lambda\sqrt{2C}t.$$

The solution is expressed in terms of hypergeometric functions

$$-\frac{1}{\xi^\delta}{}^G_2F_1\left(\frac{\delta}{2}, \frac{\delta}{2}; \frac{\delta}{2}+1; -\frac{1}{\xi^2}\right) = -\lambda\sqrt{2C}(t - t_*). \tag{2.105}$$

The latter expression gives a general relation between the *scalar field* and time.

In the special case when $\delta = 2$ (in this case $\lambda = \sqrt{3\kappa/2}/2$), it is possible to obtain the result in terms of elementary functions. The dependence of the *scalar field* on cosmic time is as follows

$$\phi(t) = 2\sqrt{\frac{2}{3\kappa}}\coth^{-1}\left[\exp\left(\frac{\sqrt{3\kappa C}}{2}(t - t_*)\right)\right]. \tag{2.106}$$

Then the *scale factor*, after integrating and the substitution of the dependance ϕ on t, takes the form

$$a = \tilde{a}_s \left(\exp\left(\frac{\sqrt{3\kappa C}}{2}(t - t_*) \right) - 1 \right)^{1/3}. \qquad (2.107)$$

The solution for the potential (2.103) for the special case $\delta = 2$ gives the same solution for the *scalar fields* (2.106). The solution for the *scale factor* is

$$a = \tilde{a}_s \left(1 - \exp\left(\frac{\sqrt{3\kappa C}}{2}(t - t_*) \right) \right)^{1/3}. \qquad (2.108)$$

We recall that the representation of the *Hubble parameter* as a function of the *scalar field* gives us the possibility to calculate the number of e-folds quite easily using the formula

$$N(\phi) = -\frac{\kappa}{2} \int_{t_i}^{t_f} \frac{H}{H'} d\phi.$$

Thus we may calculate the number of e-folds with any value of δ with the relation

$$N(\phi) = -\frac{\kappa}{2\lambda^2\delta} \ln\cosh(\lambda\phi)$$

The discussion about the solutions (2.107)–(2.108) is carried out in the work [32].

Part II

Advanced methods of exact solutions' construction

Chapter 3

Approximated methods

3.1 The standard approximations

Let us recall that the *scalar field* may be presented as a *perfect fluid* with the following relation on it's parameters

$$\rho_\phi = \frac{\dot{\phi}^2}{2} + V(\phi), \quad p_\phi = \frac{\dot{\phi}^2}{2} - V(\phi).$$

With this approach, we may slightly transform the SCEs (1.10)–(1.12) to the following form

$$3H^2 = \rho_\phi = \frac{\dot{\phi}^2}{2} + V(\phi), \tag{3.1}$$

$$2\dot{H} + 3H^2 = -p_\phi = -\frac{\dot{\phi}^2}{2} + V(\phi). \tag{3.2}$$

The evolution equation for the *scalar field*

$$\ddot{\phi} + 3H\dot{\phi} + V'(\phi) = 0, \tag{3.3}$$

is analogous to the law of energy conservation

$$\dot{\rho} + 3H(\rho + p) = 0. \tag{3.4}$$

In the following we will restrict our study to expansionary *cosmological models*, which satisfy the condition that the *scale factor* is a monotonically increasing function of time. For expanding *cosmological models*, the condition $H > 0$ is always satisfied. *Cosmological models* with $H < 0$ correspond to a contracting *universe*, in which the *scale factor* is a monotonically decreasing function of time.

3.1.1   *The inflationary attractor*

The behaviour of *inflationary attractor* was established by Salopek and
Bond [21, 34], and we now discuss its properties. To investigate the *inflationary attractor* properties we will start from the ISB presentation of scalar
field cosmology in the form:

$$V(\phi) = 3H_0^2(\phi) - 2H_0'^2(\phi), \tag{3.5}$$

$$\dot{\phi} = -2H_0'(\phi). \tag{3.6}$$

Suppose $H_0(\phi)$ is any solution to the SCEs. Let us consider a linear perturbation $\delta H(\phi)$. We assume that the perturbation does not reverse the
sign of $\dot{\phi}$. One can easily check that the perturbation obeys the linearised
equation

$$H_0'\delta H' \simeq \frac{3}{2}H_0\delta H, \delta V = 0, \tag{3.7}$$

which has the general solution

$$\delta H(\phi) = \delta H(\phi_i) \exp\left(\frac{3}{2}\int_{\phi_i}^{\phi} \frac{H_0}{H_0'}d\phi\right). \tag{3.8}$$

Since $H_0' > 0$ and $d\phi < 0$ have, from Eq. (3.6), the integrand within the
exponential term is negative definite, and hence all linear perturbations die
away.

If H_0 is inflationary, the behaviour is particularly dramatic because the
condition for *inflation* bounds the integrand away from zero. Consequently,
one obtains

$$\delta H(\phi) < \delta H(\phi_i) \exp\left(-\sqrt{\frac{3}{2}}|\phi - \phi_i|\right). \tag{3.9}$$

That is, if there is an *inflationary* solution, all linear perturbations approach
it at least exponentially fast as the *scalar field* rolls.

Another way of writing the solution for the perturbation, regardless of
whether H_0 is *inflationary* or not, is in terms of the amount of expansion
which occurs by using the number of e-foldings N. In this case, we get the
decay solution [21, 34]

$$\delta H(\phi) = \delta H(\phi_0) \exp(-3|N_i - N|). \tag{3.10}$$

Hence, all linear perturbations $\delta H(\phi)$ fade out exponentially fast as the
number of e-folds increases.

For nonlinear perturbations the problem is more complicated. Although all solutions are easily matched to one another, we have not shown that they do this exponentially fast. The most inconvenient is the case when the perturbation actually changes the sign of $\dot{\phi}$, since the ISB equations are singular when this occurs. Nevertheless, while the perturbation is not enough to knock out the *scalar field* to the maximum in the *potential*, the perturbed solution will inevitably unfold and then pass through the initial value of the *scalar field* ϕ. Then it can be regarded as a perturbation with the same sign $\dot{\phi}$ as the initial solution. Therefore, the following picture emerges. If the *potential* is capable of supporting *inflation*, then *inflationary* solutions quickly approach each other with exponential speed once in a linear mode. Even when *inflation* ends, the *universe* continues to expand, and so solutions continue to approach each other. Consequently, even the exit from inflation does not depend on the initial conditions. We note that the concept of an "attractor solution" does not exist: all solutions are attractors for one another and converge asymptotically. As one can see, this is a vital requirement if the expansion of a slow roll makes sense [21, 34].

3.2 Slow-roll approximation

Inflationary universe models are based upon the possibility of slow evolution of some scalar field ϕ in a *potential* $V(\phi)$ [35]. Although some *exact solutions* of *inflationary* models exist, most detailed studies of *inflation* have been made using numerical integration, or by employing an approximation scheme. The '*slow-roll approximation*' [21, 36, 37], which neglects the most slowly changing terms in the equations of motion, is the most widely used. Although this approximation works well in many cases, we know that it must eventually fail if *inflation* is to end. Moreover, even weak violations of *inflation* can result in significant deviations from the standard predictions for observables such as the spectrum of density perturbations or the density of *gravitational waves* in the *universe* [36, 38]. As observational data sharpen, it is important to derive a suite of predictions for the observables that are as accurate as possible, and which cover all possible *inflationary* models.

In the literature, one finds two different versions of the *slow-roll approximation*. The first [36] places restrictions on the form of the *potential*, and requires the evolution of the *scalar field* to have reached its asymptotic form. This approach is most appropriate when studying *inflation* in a specific *potential*. It is called the *potential slow-roll approximation* (V-SR).

The other form of the approximation places conditions on the evolution of the *Hubble parameter* during *inflation* [39]. It is called the *Hubble slow-roll approximation* (H-SR). It has distinct advantages over the *potential slow-roll approximation*, possessing a clearer geometrical interpretation and more convenient analytic properties. These make the H-SR approximation best suited for general studies, where the *potential* is not specified.

3.2.1 *The potential-slow-roll approximation*

When provided with a *potential* $V(\phi)$ from which to construct an *inflationary* model, the *slow-roll approximation* is normally advertised as requiring the smallness of the two parameters (both functions of ϕ), defined by [36]

$$\epsilon_V(\phi) = \frac{1}{2}\left(\frac{V'(\phi)}{V(\phi)}\right)^2, \qquad (3.11)$$

$$\eta_V(\phi) = \frac{V''(\phi)}{V(\phi)}. \qquad (3.12)$$

Henceforth, we refer to them as V-SR parameters. Their smallness is used to justify the neglect of the kinetic term in the Friedmann equation, Eq. (3.1), and the acceleration term in the scalar wave equation, Eq. (3.3). Unfortunately, the smallness of the potential-slow-roll parameters is a necessary consistency condition, but not a sufficient one to guarantee that those terms can be neglected. The *potential-slow-roll parameters* only restrict the form of the potential, not the properties of dynamic solutions. The solutions are more general because they possess a freely specifiable parameter, the value of $\dot{\phi}$, which governs the size of the kinetic term. The kinetic term could, therefore, be as large as one wants, regardless of the smallness, or otherwise, of these V-SR parameters.

In general, this *potential-slow-roll* formalism requires a further 'assumption'; that the *scalar field* evolves to approach an asymptotic attractor solution, determined by

$$\dot{\phi} \simeq -\frac{V'}{3H}. \qquad (3.13)$$

The word 'assumption' is placed in quotes here because, in general, one is able to test whether Eq. (3.13) is approached for a wide range of initial conditions.

3.2.2 *The Hubble slow-roll approximation*

If $H(\phi)$ is taken as the primary quantity, then there is a better choice of *slow-roll parameters*. We define the *Hubble-slow-roll parameters* ϵ_H and η_H by [39]

$$\epsilon_H(\phi) = 2\left(\frac{H'(\phi)}{H(\phi)}\right)^2, \tag{3.14}$$

$$\eta_H(\phi) = 2\frac{H''(\phi)}{H(\phi)}. \tag{3.15}$$

These possess an extremely useful set of properties which make them superior choices to ϵ_V and η_V as descriptors of *inflation*:

- We have exactly

$$\epsilon_H = 3\frac{\dot\phi^2/2}{V + \dot\phi^2/2} = -\frac{d\ln H}{d\ln a}, \tag{3.16}$$

$$\eta_H = -3\frac{\ddot\phi}{3H\dot\phi} = -\frac{d\ln\dot\phi}{d\ln a} = -\frac{d\ln H'}{d\ln a}. \tag{3.17}$$

- $\epsilon_H \ll 1$ is the condition for neglecting the kinetic term.
- $|\eta_H| \ll 1$ is the condition for neglecting the acceleration term.
  As a consequence, all the necessary dynamical information is encoded in the *Hubble-slow-roll parameters*. They do not need to be supplemented by any assumptions about the *inflationary attractor*, Eq. (3.13).
- The condition for *inflation* to occur is precisely

$$\ddot a > 0 \Longleftrightarrow \epsilon_H < 1. \tag{3.18}$$

There is an algebraic expression relating ϵ_V to ϵ_H and η_H:

$$\epsilon_V = \epsilon_H\left(\frac{3 - \eta_H}{3 - \epsilon_H}\right)^2. \tag{3.19}$$

The true endpoint of inflation, gauged by the *Hubble-slow-roll parameters*, occurs exactly at $\epsilon_H = 1$. When using the *potential-slow-roll parameters*, this condition is approximate; *inflation* ending at $\epsilon_V = 1$ is only a first-order result.

For η_V, the relation to the *Hubble-slow-roll parameters* is differential rather than algebraic,

$$\eta_V = \sqrt{2\epsilon_H}\frac{\eta_H'}{3 - \epsilon_H} + \left(\frac{3 - \eta_H}{3 - \epsilon_H}\right)(\epsilon_H + \eta_H). \tag{3.20}$$

This will show that the first term in Eq. (3.20) is of higher-order in slow-roll, so that to lowest-order, one has $\eta_V = \eta_H + \epsilon_H$. Note that η_H and η_V are not the same to first-order in slow-roll, as one expects from $H^2 \propto V$. We could have defined η_H to coincide with η_V in slow-roll, by defining $\bar{\eta}_H = \eta_H - \epsilon_H$, but we prefer to regard the definitions in Eqs. (3.14) and (3.15) as fundamental.

The definitions can be used to derive two useful relations between parameters of the same type

$$\eta_H = \epsilon_H - \frac{\epsilon'_H}{\sqrt{2\epsilon_H}}, \tag{3.21}$$

$$\eta_V = 2\epsilon_V - \frac{\epsilon'_V}{\sqrt{2\epsilon_V}}. \tag{3.22}$$

Note that although, as functions, the parameters of a given type, either *Hubble slow-roll* or *potential-slow-roll*, are related, their values at a given ϕ are independent of each other. One immediately sees the different 'normalisation' of the η from Eqs. (3.21) and (3.22).

3.3 The approximate solutions

In the article [34] the approximate solutions were derived by *slow-roll approximation* as:

$$\begin{aligned}
H^2(\phi) = V(\phi) \Big[& 1 + \frac{1}{3}\epsilon_V - \frac{1}{3}\epsilon_V^2 + \frac{2}{9}\epsilon_V \eta_V + \frac{25}{27}\epsilon_V^3 + \frac{5}{27}\epsilon_V \eta_V^2 - \frac{26}{27}\epsilon_V^2 \eta_V \\
& + \frac{2}{27}\epsilon_V \xi_V^2 - \frac{327}{81}\epsilon_V^4 + \frac{460}{81}\epsilon_V^3 \eta_V - \frac{172}{81}\epsilon_V^2 \eta_V^2 + \frac{14}{81}\epsilon_V \eta_V^3 - \frac{44}{81}\epsilon_V^2 \xi_V^2 \\
& + \frac{2}{9}\epsilon_V \eta_V \xi_V^2 + \frac{2}{81}\epsilon_V \sigma_V^3 + \mathcal{O}_5 \Big].
\end{aligned} \tag{3.23}$$

The general isotropic solution for a given potential possesses one free parameter, corresponding to the freedom to specify H (or equivalently $\dot{\phi}$) at some initial time. Unless an attractor exists and has been attained, there is no need for this single solution to represent in any way the true solution for that *potential*. However, if the attractor has indeed been reached, then any particular solution provides an excellent approximation to those arising from a wide range of initial conditions. This is particularly important when inflation approaches its end, and the *slow-roll parameters* become large, because one might naively assume that the one-parameter freedom could be important there. If the attractor solution exists, then solutions for a wide range of initial conditions will converge, and subsequently all exit *inflation* in the same way.

Thus, one can generate an analytic solution for *inflation* in the *potential* $V(\phi)$, that is accurate up to fourth-order in the *slow-roll parameters*, rather than the usual lowest-order.

Chapter 4

Advanced methods of exact solutions' construction

4.1 The method of generating functions

By adding Eqs. (3.1) and (3.2), we obtain a *Riccati type equation* satisfied by H, of the form

$$\dot{H} = V - 3H^2. \tag{4.1}$$

By substituting the Hubble function from Eq. (3.1) into Eq. (3.3), we obtain the basic equation describing the *scalar field* evolution as

$$\ddot{\phi} + \sqrt{\frac{3}{2}}\sqrt{\dot{\phi}^2 + 2V\left(\phi\right)}\,\dot{\phi} + \frac{dV}{d\phi} = 0. \tag{4.2}$$

Equation (4.2) can be solved by using the *generating functions* that depend on the *scalar field* and the *scale factor*.

4.1.1 The first class of generating functions

Defining the function $F = F(\phi)$ via [40]

$$\dot{\phi} \equiv \pm\sqrt{(F-1)V} \tag{4.3}$$

gives us the equation:

$$\ddot{\phi} \pm \sqrt{\frac{3}{2}}V\sqrt{F^2 - 1} + V' = 0. \tag{4.4}$$

Equations (4.3) and (4.4) form a set of coupled equations which are equivalent to Eq. (4.2). They can be uncoupled by differentiating (4.3) with respect to time and substituting for $\ddot{\phi}$ in (4.4). Solving for $\ddot{\phi} = \frac{1}{2}\frac{d\dot{\phi}^2}{d\phi}$ from (4.3) we have (with $F' \equiv \frac{dF}{d\phi}$)

$$\ddot{\phi} = \frac{1}{2}\left[(F-1)V' + VF'\right], \tag{4.5}$$

and inserting this expression for $\ddot{\phi}$ into (4.4) leads to

$$(F+1)V' + VF' \pm \sqrt{6}V\sqrt{F^2-1} = 0. \tag{4.6}$$

It can be seen that if one chooses $F \equiv F(\phi)$, then Eq. (4.6) is always separable and the *potential* is given by

$$V = \beta \exp\left(-\int \frac{F' \pm \sqrt{6}\sqrt{F^2-1}}{F+1} d\phi\right), \tag{4.7}$$

with $F \equiv F(\phi)$, and β is a constant. This may be simplified to

$$V = \frac{B}{F+1} \exp\left(\mp\sqrt{6} \int \sqrt{\frac{F-1}{F+1}} d\phi\right), \tag{4.8}$$

where B is a constant.

If one chooses $F = const$, Eq. (4.8) gives

$$V = \frac{B}{F+1} \exp\left(\mp\sqrt{6}\sqrt{\frac{F-1}{F+1}} \phi\right), \tag{4.9}$$

with solutions from (4.3) of the form

$$\phi(t) = \pm\frac{1}{\sqrt{6}}\sqrt{\frac{F+1}{F-1}} \ln\left[\pm\frac{\sqrt{6}(F-1)\sqrt{B}}{\sqrt{F+1}}(t-C)\right] \tag{4.10}$$

$$a(t) \propto t^{\frac{1}{3}\left(\frac{F+1}{F-1}\right)}. \tag{4.11}$$

Another *exact solution* can be obtained from $F = \cosh(\lambda\phi)$. Then Eq. (4.8) gives

$$V(\phi) = C\left(1 + \cosh\lambda\phi\right)^{\mp(2\sqrt{6}/\lambda)-1}, \tag{4.12}$$

$$\dot{\phi} = \pm\sqrt{C}\frac{\sinh\lambda\phi}{(1+\cosh\lambda\phi)^{1\pm g'}}. \tag{4.13}$$

This equation is only consistent if all upper or all lower signs are taken, i.e., one should not mix upper and lower signs. Chose $\lambda = \sqrt{6}$.

The upper $-$ sign in Eq. (4.12), corresponding to the upper $+$ sign in definition (4.3), gives

$$V(\phi) = \frac{C}{(1+\cosh\lambda\phi)^3}, \tag{4.14}$$

and the lower + sign in (4.12) corresponding to the lower − sign in definition (4.3), gives

$$V(\phi) = C(1 + \cosh \lambda \phi). \tag{4.15}$$

Using this result (4.15) in Eq. (4.3) leads to

$$\dot{\phi} = -\sqrt{C} \sinh \lambda \phi. \tag{4.16}$$

Solving for $\phi(t)$ yields

$$\phi(t) = \frac{2}{\lambda} \coth^{-1} \left\{ \exp \left[\lambda \sqrt{C}(t - D) \right] \right\}$$

$$= \frac{1}{\lambda} \ln \left\{ \frac{\exp \left[\lambda \sqrt{C}(t - D) \right] + 1}{\exp \left[\lambda \sqrt{C}(t - D) \right] - 1} \right\}, \tag{4.17}$$

where D is a constant. This form of the *potential* suggests that $\phi(t)$ be a function that decreases from an initial maximum value similar to the chaotic *inflation* model. We can choose $D = 0$ and so (4.17) can be written as

$$\phi(t) = \frac{1}{\lambda} \ln \left[\frac{\exp(\lambda \sqrt{C}t) + 1}{\exp(\lambda \sqrt{C}t) - 1} \right]. \tag{4.18}$$

This solution can be used to determine the evolution of the *scale factor* with time for the expanding (+ square root) solution:

$$a(t) = a_0 \left[\exp(2\lambda \sqrt{C}\, t) - 1 \right]^{\frac{1}{3}}. \tag{4.19}$$

4.1.2 The second class of generating functions

By defining a new function $f(\phi)$ so that

$$\dot{\phi} = \sqrt{f(\phi)}, \quad f(\phi) = 2V(\phi) \sinh^2 G(\phi), \tag{4.20}$$

and changing the independent variable from t to ϕ, Eq. (4.2) becomes [41]

$$\frac{dG}{d\phi} + \frac{1}{2V} \frac{dV}{d\phi} \coth G + \sqrt{\frac{3}{2}} = 0, \tag{4.21}$$

where the function G can be obtained from the *scalar field* with the use of the equation

$$G(\phi) = \operatorname{arccosh} \sqrt{1 + \frac{\dot{\phi}^2}{2V(\phi)}}. \tag{4.22}$$

We consider the case in which the *scalar field potential* can be represented as a function of G in the form

$$\frac{1}{2V}\frac{dV}{d\phi} = \sqrt{\frac{3}{2}}\,\alpha_1 \tanh G, \qquad (4.23)$$

where α_1 is an arbitrary constant. With this choice, the evolution equation takes the simple form

$$\frac{dG}{d\phi} = \sqrt{\frac{3}{2}}\,(1+\alpha_1), \qquad (4.24)$$

with the general solution given by

$$G(\phi) = \sqrt{\frac{3}{2}}\,(1+\alpha_1)\,(\phi-\phi_0), \qquad (4.25)$$

where ϕ_0 is an arbitrary constant of integration. With the use of this form of G, we obtain the self-interaction *potential* of the *scalar field* and the *scale factor*

$$V(\phi) = V_0 \cosh^{\frac{2\alpha_1}{1+\alpha_1}}\left[\sqrt{\frac{3}{2}}\,(1+\alpha_1)\,(\phi-\phi_0)\right], \qquad (4.26)$$

$$a = a_0 \sinh^{\frac{1}{3(1+\alpha_1)}}\left[\sqrt{\frac{3}{2}}\,(1+\alpha_1)\,(\phi-\phi_0)\right]. \qquad (4.27)$$

A simple solution of the gravitational field equations for a power-law type *scalar field potential* can be obtained by assuming for the function G the following form

$$G = \operatorname{arccoth}\left(\sqrt{\frac{3}{2}}\frac{\phi}{\alpha_2}\right), \qquad \text{where } \alpha_2 = constant. \qquad (4.28)$$

With this choice of G, Eq. (4.21) immediately provides the *scalar field potential* given by

$$V(\phi) = V_0 \left(\frac{\phi}{\alpha_2}\right)^{-2(\alpha_2+1)}\left[\frac{3}{2}\left(\frac{\phi}{\alpha_2}\right)^2 - 1\right], \qquad (4.29)$$

where V_0 is an arbitrary constant of integration. The time dependence of the *scalar field* is given by a simple power law,

$$\frac{\phi(t)}{\alpha_2} = \left[\frac{\sqrt{2V_0}\,(\alpha_2+2)}{\alpha_2}\right]^{\frac{1}{\alpha_2+2}}(t-t_0)^{\frac{1}{\alpha_2+2}}. \qquad (4.30)$$

The *scale factor* can be obtained from $da/d\phi = \left[(1/\sqrt{6})\coth G\right] a = (\phi/2\alpha_2)\, a$, and has an exponential dependence on the *scalar field* and the time

$$a(\phi) = a_0 \exp\left(\frac{\phi^2}{4\alpha_2}\right), \tag{4.31}$$

$$a(t) = a_0 \exp\left\{\frac{1}{4\alpha_2}\left[\frac{(\alpha_2+2)\sqrt{2V_0}}{\alpha_2}\right]^{\frac{2}{\alpha_2+2}}(t-t_0)^{\frac{2}{\alpha_2+2}}\right\}, \tag{4.32}$$

with a_0 an arbitrary constant of integration.

4.1.3   The third class of generating functions

In the paper [42] Eq. (4.1)–(4.2) are written in the following form

$$\dot{H} = V(H) - 3H^2 \tag{4.33}$$

$$\dot{\phi} = \pm\sqrt{2}\sqrt{3H^2 - V(H)}, \tag{4.34}$$

where $V(\phi) = V(\phi(t)) = V(\phi(t(H))) = V(H)$. The *potential* as a function of the *Hubble parameter* $V = V(H)$ is defined as

$$V(H) = 3H^2 + g(H) \tag{4.35}$$

and, by the choice of the *graceful exit* function $g(H)$, *exact solutions* of the system (4.1)–(4.2) are generated.

For the power–law and intermediate *inflation*

$$g(H) = -AH^n, \tag{4.36}$$

where n is real and A a positive constant. For $n = 0$, the following solutions were found:

$$H(t) = -(At + C_1) \tag{4.37}$$

$$a(t) = a_0 \exp\left(-\frac{1}{2A}(At + C_1)^2 + C_2\right) \tag{4.38}$$

$$\phi(t) = \pm\sqrt{\frac{2A}{\kappa}}(At + C_1 - C_3) \tag{4.39}$$

$$V(\phi) = 3\left(\sqrt{\frac{\phi}{2A}} + C_3\right)^2 - A. \tag{4.40}$$

For $n = 1$, we get the following solutions:

$$H(t) = C_1 \exp(-At) \tag{4.41}$$

$$a(t) = a_0 \exp\left(-\frac{C_1}{A}\exp(-At) + \frac{C_2}{A}\right) \tag{4.42}$$

$$\phi(t) = \pm\sqrt{\frac{8}{A}}\left[\sqrt{C_1}\exp\left(\frac{-At}{2} - C_3\right)\right] \tag{4.43}$$

$$V(\phi) = \frac{A}{8}e^{2C_3}\phi^2\left(\frac{3A}{8}e^{2C_3}\phi^2 - A\right). \tag{4.44}$$

For $n = 2$, we obtain the solutions:

$$H(t) = \frac{1}{At + C_1} \tag{4.45}$$

$$a(t) = a_0(C_2(At + C_1))^{1/A} \tag{4.46}$$

$$\phi(t) = \pm\sqrt{\frac{2}{A}}\ln\left(\frac{1}{C_3(At + C_1)}\right) \tag{4.47}$$

$$V(\phi) = (3 - A)C_3^2\exp(\pm\sqrt{2A}\phi). \tag{4.48}$$

For $n \neq 0, 1, 2$, we get:

$$H = (A(n - 1)(t + C_1))^{1/(1-n)} \tag{4.49}$$

$$a(t) = a_0 \exp\left[(A(n - 1))^{1/(1-n)}\frac{1 - n}{2 - n}(t + C_1)^{(2-n)/(1-n)}\right] \tag{4.50}$$

$$\phi(t) + C_3 = \sqrt{\frac{2}{A}}\frac{2}{2 - n}\left[A(n - 1)(t + C_1)\right]^{(2-n)/(2(1-n))} \tag{4.51}$$

$$V(\phi) = \sqrt{\frac{A}{8}}(2 - n)\,(\phi + C_3)^{2/(2-n)}$$

$$\times\left(3\frac{A}{8}(2 - n)^2\,(\phi + C_3)^2 - A\left(\frac{A}{8}\right)^{n/2}(2 - n)^n\,(\phi + C_3)^n\right) \tag{4.52}$$

For new *inflation* the *generating function* is considered as a polynomial in H up to second order

$$g(H) = \frac{1}{G}H^2 + \left(D - \frac{2A}{G}\right)H + \frac{A^2}{G} - AD, \tag{4.53}$$

where A, D, G are constants. The *exact solutions* in such case are:

$$H(t) = A - \frac{DG \exp(Dt + F)}{1 + \exp(Dt + F)} \tag{4.54}$$

$$a(t) = a_0 \frac{\exp(At + K)}{(1 + \exp(Dt + F))^G} \tag{4.55}$$

$$\phi(t) = \pm\sqrt{8G} \arctan\left(\exp\left(\frac{Dt + F}{2}\right)\right) + C \tag{4.56}$$

$$V(\phi) = \frac{1}{\left[1 + \tan^2\left(\pm(\phi - C)/\sqrt{8G}\right)\right]^2}$$
$$\times \left[3A^2 + (6A^2 - 6ADG - D^2G)\tan^2\left(\pm\sqrt{\frac{1}{8G}}(\phi - C)\right)\right.$$
$$\left. + 3(A - DG)^2 \tan^4\left(\pm\sqrt{\frac{1}{8G}}(\phi - C)\right)\right]. \tag{4.57}$$

For $DG = 2A$, i.e.,

$$g(H) = -2H^2/\hat{A}^2 + 2\hat{A}^2\lambda^2 \tag{4.58}$$

solutions correspond to [26]

$$\phi(t) = \hat{A}\ln[\tanh(\lambda t)], \tag{4.59}$$

$$H(t) = \hat{A}^2\lambda \coth(2\lambda t), \tag{4.60}$$

$$a(t) = a_0[\sinh(2\lambda t)]^{\hat{A}^2/2}, \tag{4.61}$$

$$V(\phi) = \hat{A}^2\lambda^2\left[(3\hat{A}^2 - 2)\cosh^2\left(\frac{\phi}{\hat{A}}\right) + 2\right]. \tag{4.62}$$

The next solution corresponds to the following choice of the *generating function*:

$$g(H) = 3\hat{A}^{-10/3}\lambda^{-2/3}6^{2/3}H^{8/3} + \hat{A}^{-2}(3\hat{A}^2 - 9)H^{6/3}$$
$$- \frac{3}{2}6^{1/3}\hat{A}^{-2/3}\lambda^{2/3}(\hat{A}^2 + 1)H^{4/3} + \frac{6^{2/3}}{12}\hat{A}^{2/3}\lambda^{4/3}H^{2/3}. \tag{4.63}$$

In the notation of [26] the solution is then

$$\phi(t) = \hat{A}\,\text{csch}(\lambda t), \tag{4.64}$$

$$H(t) = \frac{\hat{A}^2\lambda}{6}\coth^3(\lambda t), \tag{4.65}$$

$$a(t) = a_0[\sinh(2\lambda t)]^{\hat{A}^2/2}\exp\left[-\frac{\hat{A}^2}{12}\coth^2(\lambda t)\right], \tag{4.66}$$

$$V(\phi) = \frac{\lambda^2}{12\hat{A}^2}\phi^2(\phi^2 + \hat{A}^2)\left(\frac{\phi^4}{\hat{A}^4} + 2\phi^2 + \hat{A}^2 - 6\right). \tag{4.67}$$

For a deflationary scenario, the Hubble expansion rate increases during the beginning of the *universe*, and then becomes constant:

$$g(H) = \frac{AC}{1 + \tan^2(H/C)} = AC\cos^2(H/C) \tag{4.68}$$

$$H = C\arctan(At + B) \tag{4.69}$$

$$a(t) = \frac{a_0\exp\left[(Ct + BC/A)\arctan(At + B) + F\right]}{[1 + (At + B)^2]^{C/(2A)}} \tag{4.70}$$

$$\phi(t) = \sqrt{-\frac{2C}{A}}\arcsin(At + B) - D \tag{4.71}$$

$$V(\phi) = = -\frac{1}{\kappa}\left[3C^2\arctan^2\left(\sinh\left(\sqrt{-\frac{A\kappa}{2C}}(\phi + D)\right)\right)\right. \tag{4.72}$$

$$\left. + \frac{AC}{1 + \sinh^2\left(\sqrt{-\frac{A\kappa}{2C}}(\phi + D)\right)}\right]. \tag{4.73}$$

A further solution is found if one supposes that the *Hubble parameter* is not constant but is increasing logarithmically:

$$g(H) = AC\exp(-H/C) \tag{4.74}$$

$$H = C\ln(At + B) \tag{4.75}$$

$$a(t) = a_0(At + B)^{(Ct + BC/A)}e^{-(Ct + BC/A) + F}, \tag{4.76}$$

$$\phi(t) = \sqrt{-\frac{8C}{A}}\sqrt{At + B} - D, \tag{4.77}$$

$$V(\phi) = \left\{3C^2\left[\ln\left(-\frac{A}{8C}(\phi + D)^2\right)\right]^2 - 8C^2\frac{1}{(\phi + D)^2}\right\}. \tag{4.78}$$

For a real *scalar field*, the constraints $C/A < 0$ and $\sqrt{At + B} > 0$ have to be satisfied. *Inflation* starts at the time $t_i = -B/A$, with a short increase (because of the exponential function in the *scalar field*), but then decreases very rapidly to zero (because of the function of the type t^t).

4.1.4 *The fourth class of generating functions*

In the article [43] the *Hubble parameter* is taken as

$$H(\phi, \dot{\phi}) = -\frac{1}{3\dot{\phi}} \frac{dG^2(\phi)}{d\phi}, \tag{4.79}$$

where $G(\phi)$ is the *generating function.*

The dynamic equations, in terms of this function, are

$$V(\phi) = G^2(\phi) - \frac{2}{3}[G'(\phi)]^2, \tag{4.80}$$

$$\dot{\phi} = -\frac{2}{\sqrt{3}} G'(\phi), \qquad H = \frac{\dot{a}}{a} = \frac{1}{\sqrt{3}} G(\phi). \tag{4.81}$$

For a constant potential $V(\phi) = \Lambda > 0$, two *generating functions* are considered, firstly:

$$G(\phi) = \sqrt{\Lambda}. \tag{4.82}$$

This gives de Sitter space-time with the *scale factor* expanding exponentially.

$$\dot{\phi} = 0, \quad H = H_I \equiv \sqrt{\Lambda/3}. \tag{4.83}$$

The second *generating function* is given by

$$G(\phi) = \frac{e^{\sqrt{\frac{3}{2}}\phi} + \Lambda e^{-\sqrt{\frac{3}{2}}\phi}}{2}. \tag{4.84}$$

The *scalar field* and the *scale factor* behave as

$$\phi = \sqrt{\frac{2}{3}} \log\left(\sqrt{3}H_I \tanh\left(\frac{3H_I}{2}t\right)\right), \quad a(t) = a_0 \sinh^{1/3}(3H_I t). \tag{4.85}$$

For the exponential *potential* $V(\phi) = \Lambda e^{\sqrt{6}\beta\phi}$ the *generating function* has the form

$$G(\phi) = \sqrt{\frac{\Lambda}{1 - \beta^2}} e^{\sqrt{\frac{3}{2}}\beta\phi}, \tag{4.86}$$

where a real generating function of this form exists only when $|\beta| < 1$. The *scalar field* and *scale factor* corresponding to this are given by

$$\phi(t) = -\sqrt{\frac{2}{3}\frac{1}{\beta}} \log\left(1 + \beta^2 \sqrt{\frac{3\Lambda}{1-\beta^2}}t\right),$$

$$a(t) = a_0 \left(1 + \beta^2 \sqrt{\frac{3\Lambda}{1-\beta^2}}t\right)^{\frac{1}{3\beta^2}}. \tag{4.87}$$

For the power-like *potentials*, we get:

$$G(\phi) = \sqrt{\Lambda}\left(1 + \frac{\mu}{n+1}|\phi|^{n+1}\right) \tag{4.88}$$

$$V(\phi) = \Lambda\left(1 + \frac{\mu}{n+1}|\phi|^{n+1}\right)^2 - \frac{2}{3}\Lambda\mu^2\phi^{2n}. \tag{4.89}$$

$$\phi(t) = (2(n-1)\mu H_I t)^{-\frac{1}{n-1}},$$

$$a(t) = a_0 \exp\left(H_I t - \frac{1}{2(n+1)}(2(n-1)\mu H_I t)^{-\frac{2}{n-1}}\right) \tag{4.90}$$

with $\phi_0 = \phi(0) = 0$ and taking into account the stability of the solutions.
For the double well *potentials*, we have

$$G(\phi) = \sqrt{\Lambda}(\cosh\alpha\phi - \beta). \tag{4.91}$$

$$V(\phi) = \Lambda\left(1 - \frac{2\alpha^2}{3}\right)\left[\cosh\alpha\phi - \frac{\beta}{1-\frac{2}{3}\alpha^2}\right]^2 + V_0, \tag{4.92}$$

where $V_0 = \frac{2\alpha^2\Lambda}{3}\left(1 - \frac{\beta^2}{1-2\alpha^2/3}\right)$. The *scalar field* and the *scale factor* behave as

$$\phi(t) = \frac{1}{\alpha}\log\coth\left(\alpha^2 H_I t\right),$$

$$a(t) = a_0 e^{-\beta H_I t}\left[\sinh\left(2\alpha^2 H_I t\right)\right]^{\frac{1}{2\alpha^2}}, \tag{4.93}$$

where the *scalar field* rolls down the *potential* from the positive side initially.
The following choice of the *generating function*

$$G(\phi) = \sqrt{\Lambda}\left(1 + \frac{\mu}{2}\phi^2 + \frac{\alpha}{3}\phi^3\right) \tag{4.94}$$

leads to the solutions

$$V(\phi) = \Lambda\left(1 + \frac{\mu}{2}\phi^2 + \frac{\alpha}{3}\phi^3\right)^2 - \frac{2\Lambda}{3}\phi^2(\mu + \alpha\phi)^2 \tag{4.95}$$

$$\phi(t) = \frac{\mu}{Ae^{2H_I\mu t} - \alpha},$$

$$a(t) = a_0 \exp\left[\frac{12\alpha^2 H_I\left(H_I + \frac{\mu^3}{6\alpha^2}\right)t - \mu^2\left(\frac{\alpha(Ae^{2H_I\mu t} - \alpha) + \alpha^2}{(Ae^{2H_I\mu t} - \alpha)^2}\right)}{12\alpha^2 H_I}\right.$$

$$\left. + \frac{\log|Ae^{2H_I\mu t} - \alpha|}{12\alpha^2 H_I}\right], \tag{4.96}$$

where A is an integration constant. If the *scalar field* starts to evolve from the value $\phi_i > -\mu/\alpha$, it will arrive at $\phi = 0$ asymptotically, where the unverse will expand exponentially. On the other hand, if it starts from values $\phi_i < -\mu/\alpha$, it evolves to negative infinity.

4.1.5 *The fifth class of generating functions*

In the paper [44] the *potential* of the *scalar field* is considered as

$$V[\phi(a)] = \frac{F(a)}{a^6}, \tag{4.97}$$

where $F(a)$ is the *generating function*. From the *Klein-Gordon equation*, we obtain

$$\frac{1}{2}\dot{\phi}^2 + V(\phi) - \frac{6}{a^6}\int da\frac{F}{a} = \frac{C}{a^6}, \tag{4.98}$$

where C is an arbitrary integration constant. Then, the problem has been reduced to quadratures:

$$\Delta t = \sqrt{3}\int\frac{da}{a}\left[\frac{6}{a^6}\int da\frac{F}{a} + \frac{C}{a^6}\right]^{-1/2} \tag{4.99}$$

$$\Delta\phi = \sqrt{6}\int\frac{da}{a}\left[\frac{-F + 6\int daF/a + C}{6\int daF/a + C}\right]^{1/2}, \tag{4.100}$$

where $\Delta t \equiv t - t_0$, $\Delta\phi \equiv \phi - \phi_0$ and t_0, ϕ_0 are arbitrary integration constants.

Consider the *generating function*

$$F(a) = Ba^s(b + a^s)^n, \tag{4.101}$$

where $B > 0$, $b > 0$, s and n are constants and $s(n+1) = 6$. Taking $C = 0$, the *potential* is obtained as

$$V(\phi) = B\left[\cosh\left(\frac{s}{2\sqrt{6}}\Delta\phi\right)\right]^{2n}. \tag{4.102}$$

This *potential* has a nonvanishing minimum at $\Delta\phi = 0$ for $s > 0$, which is equivalent to an effective *cosmological constant*. When $s < 0$, the origin becomes a maximum, and the potential vanishes exponentially for large ϕ.

Equation (4.99) was evaluated for some values of s in this paper

$$\Delta t = \sqrt{\frac{3}{B}} \left[\operatorname{arcsinh}\left(\frac{a}{\sqrt{b}}\right) - \frac{a}{(b+a^2)^{1/2}} \right], \quad s = 2 \qquad (4.103)$$

$$a = \left\{ b \left[\exp\left(\sqrt{3B}\Delta t\right) - 1 \right] \right\}^{1/3}, \qquad s = 3. \qquad (4.104)$$

For $s > 0$, the evolution begins from a singularity as $\Delta t^{1/3}$ and is asymptotically *de Sitter* with $\Delta\phi \to 0$ for $t \to \infty$. On the other hand, for $s < 0$ the evolution has a deflationary behaviour from a *de Sitter* era in the far past to a Friedmann behavior $\Delta t^{1/3}$ when $t \to \infty$.

4.2   A Schrödinger type equation

In order to describe the dynamics of the *scalar field* during *inflation*, the usual treatment is performed [45], finally leading to the pair of equations

$$3H^2 = \frac{1}{2}\dot{\phi}^2 + V(\phi) + \Lambda, \qquad (4.105)$$

$$\ddot{\phi} + 3H\dot{\phi} = -\frac{dV(\phi)}{d\phi}. \qquad (4.106)$$

The time derivative of Eq. (4.105) is related to Eq. (4.106) through the momentum equation

$$\dot{H} = -\frac{1}{2}\dot{\phi}^2. \qquad (4.107)$$

With the use of Eqs. (4.106) and (4.107) the dynamics of the model may be described by the single equation:

$$3H^2 + \dot{H} = V(\phi) + \Lambda, \qquad (4.108)$$

which can be recognized as a *Riccati equation* for the Hubble parameter $H(t)$. In a paper [46], an ansatz was proposed to replace x for t and to assign $a(t) = \psi(x)$, leading to a nonlinear *Schrödinger equation*.

A similar approach for the case of cosmology with a *perfect fluid* has been proposed earlier, but in the context of classical mechanics [47, 48]. However, the fact that in this approach no restriction is made on the form of the *scalar field potential* allows us to probe a deeper connection between QM and *inflationary cosmology*.

By defining $\psi(t)$ through

$$H \equiv \frac{1}{3} \frac{\dot{\psi}(t)}{\psi(t)}, \tag{4.109}$$

the *Riccati equation* (4.108) can be transformed into the one dimensional *Schrödinger equation*

$$\left[-\frac{d^2}{dt^2} + 3\, V(t) \right] \psi(t) = -3\Lambda\, \psi(t). \tag{4.110}$$

We shall consider solutions to Eq. (4.110) based only on the fact that the *Hubble parameter* $H(t)$ cannot be a singular function, implying that $\psi(t)$ has to be at least a C^1 class function without zeros, but without any other restriction. For example, we may consider all ground state solutions of known exactly solvable bound state problems in QM as solutions to Eq. (4.110). Hence, an immediate equivalence arises between the SF *potential* $V(\phi)$ and the *cosmological constant* Λ, with the QM *potential* $U(x)$ and ground state energy eigenvalue E_g, respectively [46]:

$$3V(\phi(t)) + 3\Lambda \leftrightarrow 2U(x) - 2E_g. \tag{4.111}$$

This is indeed a very simple proposal, which shows that all the known exactly solvable stationary problems of 1-dimensional QM must provide at least one *exact solution* to the cosmological *Schrödinger type equation*. The general algebraic procedure is very simple: for any given QM problem, use the ground state eigenfunction $\psi_g(x)$ and Eq. (4.109) to find $H(t)$; then, use Eq. (4.107) to find $\phi(t)$, which together with Eq. (4.111) defines $V(\phi)$.

For example, in the QM case of the *simple harmonic oscillator* (SHO), where $U(x) = \omega^2 x^2/2$, the *Schrödinger equation*

$$\left[-\frac{d^2}{dx^2} + 2\left(\frac{\omega^2}{2} x^2 - E_n \right) \right] \psi_n(x) = 0 \tag{4.112}$$

has the wave functions and energy eigenvalues given by

$$\psi_n(x) = \sqrt{\frac{1}{2^n n!}} \sqrt{\frac{\omega}{\pi}}\, e^{-\frac{w}{2}x^2} H_n(\sqrt{\omega}\, x), \quad E_n = \left(n + \frac{1}{2} \right)\omega, \tag{4.113}$$

where $H_n(y)$ are the *Hermite polynomials*. In the case $n = 0$, the *Hermite polynomial* is $H_0(\sqrt{\omega}\, x) = 1$, with energy and wave function

$$E_0 = \frac{1}{2}\omega, \quad \psi_0(x) = \sqrt[4]{\frac{\omega}{\pi}}\, e^{-\frac{w}{2}x^2}. \tag{4.114}$$

To construct the corresponding cosmological variables, we replace x by t in Eq. (4.114), and use Eq. (4.109) to find the associated *Hubble parameter*

$$H(t) = -\frac{\omega}{3}t, \qquad (4.115)$$

and with the use of Eq. (4.107) we obtain the expression of the *scalar field*

$$\phi(t) = \sqrt{\frac{2w}{3}}\, t. \qquad (4.116)$$

Finally, we use Eqs. (4.115) and (4.116) to find the *potential* $V(\phi)$ and the constant Λ

$$V(\phi) = \lambda\phi^2, \quad \Lambda = -\frac{2}{3}\lambda, \qquad (4.117)$$

where $\lambda = \frac{\omega}{2}$. As one can see, the *scalar field potential* derived from the SHO *potential* turns out to be $\lambda\phi^2$. Surprisingly, one of the most useful and basic *potentials* of QM transforms into one of the most useful potentials in this cosmological model [45]. It is even more surprising that other typical QM *potentials* resemble typical *scalar field potentials* in standard *cosmology*, for example, compare the results [49] to the ones obtained in [46].

4.2.1   *QM and standard cosmology analogies*

The Schrödinger picture of standard *cosmology* seems to be a fruitful approach to the construction of *exact solutions* to the *inflationary* equations (4.105), (4.106), since all *potentials* from these known QM problems resemble known SF *potentials* . It may seem that there must exist further analogies between these two models of the micro and macro cosmos than just an algebraic resemblance.

In the present analogy, $\psi(t) = a^3(t)$ describes the way the universe volume is expanding since in the Schrödinger picture, $\psi(x)$ is related to probability conservation, hence the equivalence proposed here points to energy density conservation in this expanding *universe*. On the other hand, the only constant term in the QM problem is the energy E, which therefore determines the *cosmological constant* Λ of Eq. (4.110), which could be associated with the vacuum energy density. Therefore, the sign of Λ is completely determined by the corresponding QM problem from which the SF solution is derived, becoming an immediate check for the dynamical characteristics that one wants to determine with the proposed SF *potential*.

With respect to the scalar field $\phi(t)$, following Eqs. (4.107), (4.108) and (4.111), we can see that wherever $\dot{a}(t) \simeq 0$ [46],

$$\phi(t) \simeq \int^t dy \sqrt{2\left(E - U(y)\right)} \tag{4.118}$$

which resembles the action $S(x)$ of the quantum theory.

4.2.2 Slow Roll and WKB approximations

One further analogy deserves special attention. Beginning with the *slow-roll approximation* condition

$$\left|\frac{V'}{V}\right|^2 < 1,$$

where we should do the substitution V to $V + \Lambda$ to comply with Eq. (4.108), we can see that since $\ddot{a}/a > 0$ implies that $V + \Lambda > \dot{\phi}^2$, we have that

$$\left|\dot{V}\right|^2 = \left|\dot{\phi}V'\right|^2 < |V + \Lambda| |V'|^2 < |V + \Lambda|^3 .$$

In QM analogy, we would have to do the substitutions $\frac{dV}{dt} \rightarrow \frac{dU}{dx}$ and $|V + \Lambda| \rightarrow |E - U|$, giving [46]

$$\left|\frac{dU}{dx}\right|^2 < |E - U|^3 ,$$

which is just the WKB approximation

$$\left|\frac{d^2W}{dx^2}\right| < \left|\frac{dW}{dx}\right|^2$$

of the stationary problem, where $W(x) = \pm\sqrt{2(E - U)}$ is Hamilton's principal function.

4.3 The Ermakov–Pinney equation

In the articles [50, 51] the *Einstein equations* of gravity and the *Klein-Gordon equation* for the *scalar field* were considered as

$$H^2 + \frac{c}{a^2} = \frac{1}{3}\left[\frac{1}{2}\dot{\phi}^2 + V(\phi) + \frac{D}{a^n}\right] \tag{4.119}$$

$$\ddot{\phi} + 3H\dot{\phi} + \frac{dV}{d\phi} = 0. \tag{4.120}$$

The term D/a^n is the density of matter for the barotropic fluid with equation of state $p_{mat} = (n-3)\rho_{mat}/3$; $D \geq 0$, $0 \leq n \leq 6$. The parameter c can take the values $-1, 0$ and 1 depending on the curvature of the hypersurface $t = const$.

By using a further differentiation of (4.119) and the substitution $b = a^{n/2}$ [51] was obtained:

$$\frac{2}{n} \left(\frac{\ddot{b}}{b} - \frac{\dot{b}^2}{b^2} \right) - \frac{c}{b^{4/n}} = -\frac{1}{2} \left[\dot{\phi}^2 + \frac{nD}{3b^2} \right]. \tag{4.121}$$

Also, a new comoving time τ, such that $\dot{\tau} = b$ was defined, then Eq. (4.121) becomes

$$\frac{d^2b}{d\tau^2} + \frac{n}{4} \left(\frac{d\phi}{d\tau} \right)^2 b = -\frac{n^2 D}{12b^3} + \frac{nc}{2b^{\frac{4+n}{n}}}. \tag{4.122}$$

The latter is a remarkable example of a nonlinear yet integrable ordinary differential equation (ODE) of the form:

$$Y'' + Q(\tau)Y = \frac{\lambda}{Y^3}. \tag{4.123}$$

The particularly appealing feature of this nonlinear ODE is that its general solution can be obtained, provided that one is able to solve the linear *Schrödinger equation* $Y'' + Q(\tau)Y = 0$.

If the linearly independent solutions of the the linear *Schrödinger equation* are $Y_1(\tau)$ and $Y_2(\tau)$, then the most general possible solution of the *Ermakov-Pinney equation* is given by

$$Y(\tau) = \left(AY_1^2 + BY_2^2 + 2CY_1Y_2 \right)^{1/2}, \tag{4.124}$$

where A, B and C are constants connected through

$$AB - C^2 = \frac{\lambda}{W^2} \tag{4.125}$$

and the Wronskian $W = Y_1Y_2' - Y_2Y_1'$.

The result of (4.122) can be used conversely for constructing solutions of scalar field *cosmologies*, using the *Ermakov-Pinney equation* structure and solutions. The converse result can be proved in the following form: given Q and $\lambda = -\kappa^2 n^2 D/12 < 0$, let $Y > 0$ be a solution of

$$\frac{d^2Y}{d\tau^2} + QY = \frac{\lambda}{Y^3} + \frac{nc}{2Y^{1+\frac{4}{n}}}. \tag{4.126}$$

Define a new time coordinate t such that $\dot{\tau} = Y(\tau(t))$ and $a = Y^{2/n}$, as well as a new field ϕ satisfying:

$$\frac{n\kappa^2}{4}\left(\frac{d\phi}{d\tau}\right)^2 = Q, \tag{4.127}$$

with $Q \neq 0$. Finally, define a potential:

$$V(\phi) = \frac{12}{\kappa^2 n^2}\left(\frac{dY}{d\tau}\right)^2 - \frac{Y^2}{2}\left(\frac{d\phi}{d\tau}\right)^2 - \frac{D}{Y^2} + \frac{3c}{\kappa^2 Y^{4/n}}. \tag{4.128}$$

The $(a(\tau(t)), \phi(\tau(t)), V(\phi))$ satisfies (4.119)–(4.120).

4.3.1   Flat FRW metric and massless scalar field

In the absence of matter, $D = 0$, $\lambda = 0$ and $n = 2$. For $\lambda = 0$, and for the flat FRW metric $c = 0$, Eq. (4.126) becomes

$$\frac{d^2 Y}{d\tau^2} + QY = 0. \tag{4.129}$$

Assuming the case $V(\phi) = m^2\phi^2/2$, the particular scenario of a massless *scalar field* yields $V(\phi) = 0$, hence (4.128) becomes:

$$\frac{d^2 Y}{d\tau^2} + \frac{3}{Y}\left(\frac{dY}{d\tau}\right)^2 = 0, \tag{4.130}$$

whose solution yields:

$$Y(\tau) = A\tau^{1/4}; \tag{4.131}$$

hence from Eq. (4.129):

$$Q(\tau) = \frac{3}{16}\frac{1}{\tau^2}. \tag{4.132}$$

This, in turn, through the definition of $\dot{\tau}$, yields:

$$\tau(t) = \left(\frac{3}{4}\right)^{\frac{4}{3}} A^{4/3} t^{\frac{4}{3}} \tag{4.133}$$

and, thus, finally from (4.127):

$$\phi(t) = \sqrt{\frac{2}{3}}\log\left(\frac{3A}{4}\right) + \sqrt{\frac{2}{3}}\log(t - t_0). \tag{4.134}$$

4.3.2 *Non-flat FRW metric coupled with scalar field*

For the non-flat FRW metric with $c \neq 0$, the example of $n = 2$ and $D > c/3$ (which implies that $\lambda + c < 0$) was considered. In this case, Eq. (4.126) becomes

$$\frac{d^2Y}{d\tau^2} + QY = \frac{\lambda + c}{Y^3} \tag{4.135}$$

and the solution is

$$Y(\tau) = \left(A\tau^{\frac{3}{4}} + B\tau^{\frac{1}{4}} + 2C\tau\right)^{\frac{1}{2}}, \tag{4.136}$$

where $AB - C^2 = 4(\lambda + c)$. For $A = B = 0$, $C = 2\sqrt{|\lambda + c|}$.

$$Y(\tau) = (2C\tau)^{\frac{1}{2}}; \tag{4.137}$$

$$\phi(\tau) = \frac{\sqrt{6}}{4}\log(\tau). \tag{4.138}$$

From the definition of $\dot{\tau}$, we get

$$\tau = \frac{C}{2}(t - t_0)^2. \tag{4.139}$$

$$V(\phi) = \frac{S}{\tau} \equiv Se^{-\frac{4}{\sqrt{6}}\phi}, \tag{4.140}$$

where $S = 9c/8 - D/(2c) + 3/2$.

4.3.3 *Quadratic potential*

Matter is absent, hence $D = \lambda = 0$, for $n = 4$ and $c \geq 0$, $Y(\tau) = 2B\tau$. Then

$$Q(\tau) = \frac{c}{4B^3\tau^3}. \tag{4.141}$$

From the definition of $\dot{\tau}$, we get

$$\tau(t) = \frac{A^2}{2B}e^{2Bt}, \tag{4.142}$$

where, without loss of generality, A is a positive constant. Then, from Eq. (4.127), we find

$$\phi(t) = -\frac{\sqrt{2c}}{AB}e^{-Bt} + \alpha, \tag{4.143}$$

where α is an arbitrary constant. The potential

$$V(\phi) = 3B^2 + B^2(\phi - \alpha)^2. \tag{4.144}$$

This solution was found by means of a different approach in [18].

4.3.4 *Constant scalar field*

For $Q(\tau) = 0$ and $n = 2$, the *Ermakov-Pinney equation* has the straightforward general solution

$$Y(\tau) = \left(A\tau^2 + B + 2C\tau\right)^{1/2}, \tag{4.145}$$

with

$$AB - C^2 = \lambda + c = -\frac{\kappa^2 D}{3} + c \equiv \tilde{\lambda}. \tag{4.146}$$

Consequently,

$$\tau(t) = \frac{D}{4A} \exp\left(A^{1/2}t\right) + \frac{\tilde{\lambda}}{DA^{1/2}} \exp\left(-A^{1/2}t\right) - \frac{C}{A}. \tag{4.147}$$

The *scalar field* ϕ and the *potential* V are constants, while the *scale factor*

$$a(t) = a(0) \cosh\left(A^{1/2}t\right) + \sqrt{a(0)^2 - \frac{\tilde{\lambda}}{A}} \sinh\left(A^{1/2}t\right) \tag{4.148}$$

with

$$a(0) = \frac{D}{4A^{1/2}} + \frac{\tilde{\lambda}}{DA^{1/2}}. \tag{4.149}$$

Thus, the equations of *scalar field* dynamics can be related to the simplest Ermakov system, thus reducing the problem of solving these equations to solving a single second-order, linear ODE. The nature of the *Ermakov-Pinney equation* implies that there exists a correspondence between a spatially flat FRW *universe* containing a *scalar field* and a *cosmology* containing both a *scalar field* and a *perfect fluid* [50].

4.4 The Riccati equation

In the article [52], the evolution equation for the *scalar field* is considered as a *Riccati type equation*

$$\dot{H}(t) = V(t) - 3H^2(t), \tag{4.150}$$

which arises by adding Eqs. (3.1) and (3.2). Further, in this article, cases of dependence of H and V on time are considered. When H has the general form

$$H(t) = \alpha V(t) + \frac{1}{3\beta} g(t), \tag{4.151}$$

Eq. (4.150) can be written as

$$\alpha \dot{V} + \frac{1}{3\beta} \dot{g} = V - 3\alpha^2 V^2 - 2\frac{\alpha}{\beta} gV - \frac{1}{3\beta^2} g^2. \tag{4.152}$$

In the case $g(t) = \beta/t$, the *scalar field potential* satisfies the *Bernoulli differential equation*

$$\dot{V} = \left(\frac{1}{\alpha} - \frac{2}{t}\right) V - 3\alpha V^2. \tag{4.153}$$

For the potential

$$V(t) = \frac{e^{t/\alpha}}{t \left[V_0 t + 3t \text{Ei}\left(\frac{t}{\alpha}\right) - 3\alpha e^{t/\alpha}\right]}, \tag{4.154}$$

where $Ei(z) = -\int_{-z}^{\infty} e^{-t} dt/t$ and V_0 is an arbitrary constant of integration, the *Hubble parameter* is obtained as

$$H(t) = \left[3t - \frac{9\alpha e^{t/\alpha}}{V_0 + 3\text{Ei}\left(\frac{t}{\alpha}\right)}\right]^{-1}. \tag{4.155}$$

The *scale factor* is obtained as

$$a(t) = a_0 \left\{ 3e^{t/\alpha} - \frac{t \left[V_0 + 3\text{Ei}\left(\frac{t}{\alpha}\right)\right]}{\alpha} \right\}^{1/3}. \tag{4.156}$$

And the *scalar field* as

$$\phi(t) - \phi_0 = \pm\sqrt{2} \int \sqrt{3H^2(t) - V(t)} dt. \tag{4.157}$$

As a second example of an *exact solution*, the authors consider the case in which the *potential* satisfies the *Bernoulli differential equation*

$$\dot{V} = \frac{1}{\alpha} V - 3\alpha V^2. \tag{4.158}$$

Hence the time dependence of the *scalar field potential* can be obtained as

$$V(t) = \frac{e^{t/\alpha}}{V_0 + 3\alpha^2 e^{t/\alpha}}, \tag{4.159}$$

where V_0 is an arbitrary constant of integration. Then the function $g(t)$ must satisfy the following *Bernoulli differential equation* ,

$$\frac{1}{3}\dot{g} + 2\alpha \frac{e^{t/\alpha}}{V_0 + 3\alpha^2 e^{t/\alpha}} g + \frac{1}{3\beta} g^2 = 0. \qquad (4.160)$$

The *Hubble parameter* $H(t)$ is obtained as

$$H(t) = \frac{1}{3} \left\{ \alpha + \frac{1}{\left[\beta V_0^2 g_0 - \alpha \ln \left| V_0 + 3\alpha^2 e^{t/\alpha} \right| + t\right]^{-1} + 3\left(\alpha/V_0\right) e^{\frac{t}{\alpha}}} \right\}^{-1}. \qquad (4.161)$$

The *scale factor* is given by

$$a(t) = a_0 \left[3\alpha^2 e^{t/\alpha} \left(\beta V_0^2 g_0 + t\right) - \alpha \left(V_0 + 3\alpha^2 e^{t/\alpha}\right) \right.$$

$$\left. \times \ln \left| V_0 + 3\alpha^2 e^{t/\alpha} \right| + V_0 \left(\alpha + \beta V_0^2 g_0 + t\right) \right]^{\frac{1}{3}}, \qquad (4.162)$$

where $a_0 = 1/V_0$. The time variation of the *scalar field* can be obtained from the equation $\phi(t) - \phi_0 = \pm\sqrt{2} \int^t \sqrt{3H^2(\xi) - V(\xi)}d\xi$, and the dependence of the *potential* on the *scalar field* is obtained in parametric form.

We now consider the *potential* to satisfy the integral condition

$$V(t) = f_1(t) + 3 \left[\int^t f_1(\xi) \, d\xi + V_1 \right]^2, \qquad (4.163)$$

where a new arbitrary function $f_1(t) \in C^\infty(I)$ is defined on a real interval $I \subseteq$ Re and $V_1 \in$ Re is an arbitrary constant. By inserting (4.163) into (4.1), the latter takes the form

$$\frac{dH}{dt} = f_1(t) + 3 \left[\int^t f_1(\xi) \, d\xi + V_1 \right]^2 - 3H^2. \qquad (4.164)$$

For the *scalar field potential* (4.163), the general solution of the *Riccati equation* (4.1) is given by

$$H(t) = \frac{e^{-6V_1 t - 6 \int^t \int^\psi f_1(\xi) d\xi d\psi}}{C_1 + 3 \int^t e^{-6V_1 \eta - 6 \int^\eta \int^\psi f_1(\xi) d\xi d\psi} d\eta}$$

$$+ \int^t f_1(\xi) \, d\xi + V_1, \qquad (4.165)$$

where C_1 is an arbitrary constant of integration. The *scale factor* is

$$a\left(t\right) = a_0 e^{V_1 t + \int^t \int^\zeta f_1(\xi) d\xi d\zeta}$$

$$\times \left[C_1 + 3 \int^t e^{-6V_1\eta - 6 \int^\eta \int^\psi f_1(\xi) d\xi d\psi} d\eta \right]^{1/3}, \qquad (4.166)$$

where a_0 is an arbitrary constant of integration. The *scalar field* $\phi\left(t\right)$ can be written as

$$\phi\left(t\right) = \phi_0 \pm \sqrt{2} \int^t \left\{ -f_1(\zeta) - 3 \left[\int^\zeta f_1(\xi) d\xi + V_1 \right]^2 \right.$$

$$+ 3 \left[\frac{e^{-6V_1\zeta - 6 \int^\zeta \int^\psi f_1(\xi) d\xi d\psi}}{C_1 + 3 \int^\zeta e^{-6V_1\eta - 6 \int^\eta \int^\psi f_1(\xi) d\xi d\psi} d\eta} \right.$$

$$\left. \left. + \int^\zeta f_1\left(\xi\right) d\xi + V_1 \right]^2 \right\}^{\frac{1}{2}} d\zeta, \qquad (4.167)$$

where $\phi_{0\pm}$ are arbitrary constants of integration.

For the particular case $f_1(t) = f_0 = \text{constant} > 0$, $V_1 = 0$ and the *scalar field potential* is given by

$$V(t) = 3f_0^2 t^2 + f_0. \qquad (4.168)$$

The *Hubble parameter*

$$H(t) = \frac{2\sqrt{f_0} e^{-3f_0 t^2}}{2H_1 \sqrt{f_0} + \sqrt{3\pi} \text{erf}\left(\sqrt{3f_0}t\right)} + f_0 t, \qquad (4.169)$$

where H_1 is an arbitrary constant of integration, and $\text{erf}(z) = (2/\sqrt{\pi}) \int_0^z \exp\left(-t^2\right) dt$. The *scale factor* $a(t)$ can be obtained as

$$a(t) = a_0 e^{\frac{f_0 t^2}{2}} \left[H_1 + \frac{1}{2}\sqrt{\frac{3\pi}{f_0}} \text{erf}\left(\sqrt{3f_0}t\right) \right]^{1/3}. \qquad (4.170)$$

The *scalar field* can be obtained in an integral form as

$$\phi(t) = \phi_0 + \sqrt{2} \int^t \left[-3f_0^2 \zeta^2 - 3f_0 \right.$$

$$\left. \times \left(\frac{2\sqrt{f_0} e^{-3f_0\zeta^2}}{2H_1\sqrt{f_0} + \sqrt{3\pi}\text{erf}\left(\sqrt{3f_0}\zeta\right)} + f_0\zeta \right)^2 \right]^{1/2} d\zeta. \qquad (4.171)$$

Equations (4.168) and (4.167) give the functional dependence of the *scalar field potential* V of the *scalar field* ϕ in parametric form, with t as parameter.

When the *potential* $V_\pm(t)$ satisfies the differential condition

$$V_\pm(t) = 3f_2(t) \pm \frac{d}{dt}\sqrt{f_2(t)}, \tag{4.172}$$

where $f_2(t) \in C^\infty(I)$ is a new arbitrary function defined on a real interval $I \subseteq \mathrm{Re}$, then

$$\frac{dH_\pm}{dt} = 3f_2(t) \pm \frac{d}{dt}\sqrt{f_2(t)} - 3H_\pm^2. \tag{4.173}$$

For the scalar field potential (4.172), the general solution of the *Riccati equation* is

$$H_\pm(t) = \frac{e^{\mp 6\int^t \sqrt{f_2(\psi)}d\psi}}{C_{1\pm} + 3\int^t e^{\mp 6\int^\eta \sqrt{f_2(\psi)}d\psi}d\eta} \pm \sqrt{f_2(t)}, \tag{4.174}$$

where $C_{1\pm}$ are arbitrary constants of integration.

The *scale factor* is given by

$$a_\pm(t) = a_{0\pm}e^{\pm\int^t \sqrt{f_2(\zeta)}d\zeta}$$
$$\times \left[C_{1\pm} + 3\int^t e^{\mp 6\int^\eta \sqrt{f_2(\psi)}d\psi}d\eta\right]^{1/3}, \tag{4.175}$$

where $a_{0\pm}$ are arbitrary constants of integration. The *scalar field* $\phi(t)$ can be written as

$$\phi_\pm(t) = \phi_{0\pm} \pm \sqrt{2}\int^t \left\{-3f_2(\zeta) \mp \frac{d}{d\zeta}\sqrt{f_2(\zeta)}\right.$$
$$\left. + 3\left[\frac{e^{\mp 6\int^\zeta \sqrt{f_2(\psi)}d\psi}}{C_{1\pm} + 3\int^\zeta e^{\mp 6\int^\eta \sqrt{f_2(\psi)}d\psi}d\eta} \pm \sqrt{f_2(\zeta)}\right]^2\right\}^{\frac{1}{2}}d\zeta, \tag{4.176}$$

where $\phi_{0\pm}$ are arbitrary constants of integration.

For $f_2(t)$ having the form $f_2(t) = f_{02}/t^2$, with $f_{02} = \text{constant} > 0$, the *scalar field potential* takes the form

$$V_\pm(t) = \frac{V_{0\pm}}{t^2}, \tag{4.177}$$

where $V_{0\pm} = 3f_{02} \mp \sqrt{f_{02}}$. The *Hubble parameter*

$$H_\pm(t) = \frac{1}{6t}\left[1 + V_{1\pm}\left(1 - \frac{2H_{1\pm}}{H_{1\pm} + t^{V_{1\pm}}}\right)\right], \tag{4.178}$$

where $H_{1\pm}$ are arbitrary constants of integration, and $V_{1\pm} = \sqrt{12V_{0\pm} + 1}$. The *scale factor* is:

$$a_\pm(t) = a_{0\pm} t^{(1-V_{1\pm})/6} \left(H_{1\pm} + t^{V_{1\pm}} \right)^{1/3}, \qquad (4.179)$$

where $a_{0\pm}$ are arbitrary constants of integration. The *scalar field* $\phi(t)$ can be written as

$$\phi_\pm(t) = \phi_{0\pm} \pm \frac{1}{\sqrt{6}}$$

$$\times \int^t \frac{d\zeta}{\zeta} \sqrt{\left[V_{1\pm} \left(\frac{2H_{1\pm}}{H_{1\pm} + \zeta^{V_{1\pm}}} - 1 \right) - 1 \right]^2 - 12V_{0\pm}}. \quad (4.180)$$

In the case of $V_{0+} = -1/12$, corresponding to the value $f_{02} = 1/36$, the authors of [52] obtain a complete particular solution of the gravitational field equations describing the time evolution of the flat FRW *universe* with the self interaction *potential* $V_+(\phi)$, given by

$$H_+(t) = \frac{1}{6t}, \quad a_+(t) = a_{0+} t^{1/6}, \quad q_+ = 5,$$

$$\phi_+(t) = \phi_{0+} + \frac{\ln|t|}{\sqrt{3}}, \quad V_+(\phi) = V_{2+} e^{-2\sqrt{3}\phi}, \qquad (4.181)$$

where $V_{2+} = -e^{2\sqrt{3}\phi_{0+}}/12$. Thus, the power law solution for the *cosmological model* was obtained, with the *potential* expressed as an exponential function of the *scalar field*.

4.5   The superpotential method

The *superpotential* method for the standard *inflationary* model (1.1) was successfully applied for solving the SCEs (1.10)–(1.12). The main idea was [53] to represent the SCEs in the form of the *slow-roll approximation* , i.e., in the equations for the spatially flat *universe*,

$$H^2 = \frac{\kappa}{3} \left(\frac{1}{2} \dot\phi^2 + V(\phi) \right), \qquad (4.182)$$

$$\dot H = -\kappa \frac{1}{2} \dot\phi^2, \qquad (4.183)$$

$$\ddot\phi + 3H\dot\phi + V'(\phi) = 0, \qquad (4.184)$$

we should omit $\ddot{\phi}$, $\dot{\phi}^2$. After this, the SCEs in the *slow-roll regime* take the form

$$H^2 \simeq \frac{\kappa}{3} V(\phi) \tag{4.185}$$

$$\dot{H} = -\kappa \frac{1}{2} \dot{\phi}^2 \tag{4.186}$$

$$3H\dot{\phi} \simeq -V'(\phi). \tag{4.187}$$

To obtain the desired form of the equations, the *potential* of total energy [54] as the sum of kinetic energy (in terms of *scalar field* argument $U(\phi) = \dot{\phi}$) and the potential energy, was introduced

$$W(\phi) = \frac{1}{2} U^2(\phi) + V(\phi), \quad U(\phi) = \dot{\phi}. \tag{4.188}$$

After this substitution, the SCEs take the following form

$$H^2 = \frac{\kappa}{3} W(\phi) \tag{4.189}$$

$$\dot{H} = -\frac{\kappa}{2} U(\phi)^2 \tag{4.190}$$

$$3H\dot{\phi} = -W'(\phi). \tag{4.191}$$

As we can see, the system of equations above exactly reproduce the SCEs in *slow-roll form* (4.185)–(4.187) if we substitute $W(\phi)$ instead of $V(\phi)$.

As we know, any one from of the presented exact SCEs in the *slow-roll form* (4.189)–(4.191) can be derived as a consequence (or a differential consequence) of the other two. In our approach, we exclude from consideration Eq. (4.190) in the first step. Excluding the *Hubble parameter H* from (4.189), and inserting it into (4.191), taking into account the *superpotential* definition (4.188), we have the consequence of (4.189), (4.191) in the form

$$\sqrt{3\kappa} U^2 W^{1/2} = -W'. \tag{4.192}$$

Integrating (4.192) with respect to W, we obtain the relation:

$$W = \frac{3\kappa}{4} \left(\int U(\phi) d\phi \right)^2, \tag{4.193}$$

which leads to a new method of the construction of *exact solutions* in cosmology, viz., by suggesting that the evolution of the *scalar field* is given, one can determine the superpotential by solving the integral on the right hand side of (4.193) $\int \dot{\phi}^2 dt$. Knowing W, one can find H from the *Friedmann equation* (4.189) with the following relation

$$H = \frac{\kappa}{4}\sqrt{3\kappa}\left(\int U(\phi)d\phi\right).$$ (4.194)

Then by integration one can find the scale factor $a(t)$.

Thus the proposed method presents some combination of the two methods: *slow-roll like* presentation of exact equations [54] and obtaining cosmological solutions for the given *scalar field* evolution [26]. The advantage of the proposed method lies in the essential simplification of the integration procedure: one needs to calculate only one integral for obtaining a *superpotential* and the *Hubble parameter*. Then the *potential V* (4.188) as well as the *scale factor* $a(t)$ are calculated from the related definitions. This exhibition of procedure simplification and its effectiveness can be found in applications in *cosmology*, on the brane, in phantom and tachyon fields. The latter two have very restricted numbers of exact solutions which can be essentially extended by virtue to the *superpotential* method.

We can have a look at the *superpotential* method from another position. The system (4.182)–(4.184) has three unknowns, $\phi(t)$, $V(\phi)$ and $H(t)$ (or $a(t)$). To solve it, one of these variables has to be given *a priori*. It is customary to look for the solution for a given $V(\phi)$, but as it is known, it is very difficult to solve the SCEs for a given potential exactly.

In the *superpotential* approach, it is proposed to reduce the equations to a simpler form which helps us to solve them exactly. In order to do this, let us consider the *superpotential* function $W(\phi)$ defined in (4.188):

$$W(\phi) = V(\phi) + \frac{1}{2}U(\phi)^2. \quad U(\phi) = \dot{\phi}.$$

Now, with the change of variable $dt = d\phi/\dot{\phi}$ and using a reverse transformation from $U(\phi)$ to $\dot{\phi}$, one can obtain

$$\frac{dW}{d\phi} = \frac{dV}{d\phi} + \ddot{\phi}.$$ (4.195)

Hence, the SCEs (4.182)–(4.184) can be rewritten in the *slow-roll form* (4.185)–(4.187). Therefore we are hopeful to solve SCEs in the *superpotential* presentation because the slow roll approximation has been intensively studied.

The *superpotential* $W(\phi)$ (> 0) shows up as the main part of the *potential* function, driving the dynamics of the *Hubble parameter* H or the scale factor. To solve for them, note that Eq. (4.189) defines $\dot{a}/a$ as a function of ϕ, $H(\phi)$, which when inserted into Eq. (4.191), gives the *scalar field* $\phi(t)$ as a function of t, at least in quadratures

$$-3H(\phi)\left(\frac{dW}{d\phi}\right)^{-1}d\phi = dt. \tag{4.196}$$

Finally, inserting $\phi(t)$ into Eqs. (4.188) and (4.191) gives $V(\phi)$ and $a(t)$, respectively, and the solution is completed.

Obviously, one could simply have begun by giving $H=H(\phi)$, but it is usually desired to have some description of the *potential* instead, and for this reason it is preferable to give $W(\phi)$. One could also use $H(t)$ to determine $\phi(t)$, since

$$\frac{1}{2}\dot{\phi}^2 = -\frac{dH(t)}{dt} \tag{4.197}$$

implies that

$$\Delta\phi(t) = \pm\int\sqrt{-2\frac{dH(t)}{dt}}\,dt \tag{4.198}$$

and since $W = 3H^2(t)$, a complete knowledge of $H(t)$ fully determines the solution to the problem.

4.5.1 Examples of the exact solutions from the method of superpotential

We consider the *superpotential* $W = \lambda\phi^2$. In this case, $H^2=\lambda\phi^2/3$, and $dV_a/d\phi=2\lambda\phi$. Therefore, from Eq. (4.196) it is easy to find that

$$\Delta\phi(t) = \pm2\sqrt{\frac{\lambda}{3}}\,\Delta t \tag{4.199}$$

Hence, $\dot{\phi}$ is constant. Letting $\phi_0 = \phi(t_0 = 0) = 0$, and using Eqs. (4.188) and (4.189), we get

$$V(\phi) = \lambda\phi^2 - 2\frac{\lambda}{3}, \tag{4.200}$$

$$a(t) = a_0 e^{-\frac{\lambda}{3}t^2}. \tag{4.201}$$

Obviously, one would be tempted to pick $\lambda < 0$ in order to make $a(t)$ a growing function of t, but that would make $\phi(t)$ an imaginary function of t.

For the another type of *superpotential* $W = \lambda\phi^4$, the solution is

$$\phi(t) = \phi_0 e^{\pm 4\sqrt{\frac{\lambda}{3}}t} \tag{4.202}$$

$$V(\phi) = \lambda\phi^4 - \frac{8}{3}\lambda\phi^2 \tag{4.203}$$

$$a(t) = a_0 \exp\left\{-\frac{\phi_0^2}{8}e^{\pm 8\sqrt{\frac{\lambda}{3}}(t-t_0)}\right\}. \tag{4.204}$$

Notice the appearance of a double exponential in $a(t)$. In this case, the radiation era happens when $\phi = 2$, and *inflation* comes when $\phi > 2\sqrt{2}$.

The cases with integer $n > 2$, $W = \lambda\phi^{2n}$, all have the same kind of solution. Although this is not a very common *potential* in the literature [55], it is solved here to compare with the two previous cases.

$$\phi(t) = \left[\phi_0^{2-n} \pm 2n(n-2)\sqrt{\frac{\lambda}{3}}\,(t-t_0)\right]^{\frac{-1}{n-2}} \tag{4.205}$$

$$V(\phi) = \lambda\phi^{2n} - \lambda\frac{2n^2}{3}\phi^{2(n-1)} \tag{4.206}$$

$$a(t) = a_0 \exp\left\{\frac{-1}{4n}\left[\phi_0^{2-n} \pm 2n(n-2)\sqrt{\frac{\lambda}{3}}(t-t_0)\right]^{\frac{-2}{n-2}}\right\} \tag{4.207}$$

One of the few problems exactly solved in the literature, is that of an exponential function $W = V_o e^{-\alpha\phi}$ [56]. In this case, the solution found with the method described above gives, for $\phi_0 = 0$,

$$\phi(t) = \frac{2}{\alpha}\ln\left(1 + \frac{\alpha^2}{2}\sqrt{\frac{V_o}{3}}\,t\right) \tag{4.208}$$

$$V(\phi) = V_o\left(1 - \frac{\alpha^2}{6}\right)e^{-\alpha\phi} \tag{4.209}$$

$$a(t) = a_0\left(1 + \frac{\alpha^2}{2}\sqrt{\frac{V_o}{3}}t\right)^{\frac{2}{\alpha^2}}. \tag{4.210}$$

The reader can recognize now that this is the solution used in the literature, where the two terms of $V(\phi)$ are grouped together, leaving the essence of the competing term W out of the picture.

Another *potential* used in the literature is that of a hyperbolic cosine $W = V_o\left(\cosh(\beta\phi) - 1\right)$ [57]. The solution found here, with $\phi_0 = 0$, is

$$\phi(t) = \frac{-2}{\beta}\operatorname{arcsinh}\left(\tan\left(\sqrt{\frac{V_o}{6}}\beta^2 t\right)\right) \tag{4.211}$$

$$V(\phi) = V_o\left(\cosh(\beta\phi) - 1\right) - V_o\frac{\beta^2}{6}\left(\cosh(\beta\phi) + 1\right) \tag{4.212}$$

$$a(t) = a_0 \cos^{\frac{2}{\beta^2}}\left(\sqrt{\frac{V_o}{6}}\beta^2 t\right). \tag{4.213}$$

The previous examples were related to *potentials* used in the literature. They may, however, not be able to comply with the desired characteristics of the model. Nevertheless, one can now generate a variety of solutions to (4.189)–(4.190) that may be more useful for that purpose.

We consider the *superpotential* [58]

$$W = \alpha^2 \left(e^{\beta\phi} - 1\right)^2. \tag{4.214}$$

Then, we will have that

$$\phi(t) = -\frac{1}{\beta}\ln\left(1 \pm \frac{2\beta^2\alpha}{\sqrt{3}}t\right) \tag{4.215}$$

and

$$V(\phi) = \alpha^2\left[\left(1 - \frac{2\beta^2}{3}\right)e^{2\beta\phi} - 2e^{\beta\phi} + 1\right]. \tag{4.216}$$

By requiring the lower sign in the previous equations, we find that $a(t)$ is

$$a(t) = a_0\, e^{\alpha t/\sqrt{3}}\left(1 - \frac{2\beta^2\alpha}{\sqrt{3}}\right)^{\sqrt{3}/2\beta^2\alpha}. \tag{4.217}$$

4.6 Determination of the difference between approximate and exact solutions

We write the function of the difference between approximate and exact solutions in the *e-folds number* Δ_N based on the *superpotential* method as follows [59]

$$\Delta_N(\phi) = \int\left(\frac{W}{W'} - \frac{V}{V'}\right)d\phi. \tag{4.218}$$

Table 4.1   Difference between approximate and exact solutions.

Exact superpotential W and potential V of a scalar field ϕ	Difference in the number of e-folds between approximate and exact solutions Δ_N
$W(\phi) = \lambda e^{-\alpha\phi}$ $V(\phi) = \lambda\left(1 - \frac{\alpha^2}{6}\right)e^{-\alpha\phi}$	$\Delta_N(\phi) = 0$ $\Delta_N(t) = 0$
$W(\phi) = \lambda\phi^2$ $V(\phi) = \lambda\phi^2 - \frac{2}{3}\lambda$	$\Delta_N(\phi) = \frac{1}{3}\ln(\phi), \qquad \phi_0 = 1$ $\Delta_N(t) = \frac{1}{3}\ln\left[\pm 2\sqrt{\frac{\lambda}{3}}\, t + \phi_0\right]$
$W(\phi) = \lambda\phi^{-2}$ $V(\phi) = -\frac{2}{3}\lambda\phi^{-4} + \lambda\phi^{-2}$	$\Delta_N(\phi) = \frac{1}{6}\ln\left(3\phi^2 - 4\right), \qquad \phi_0 = \sqrt{5/3}$ $\Delta_N(t) = \frac{1}{6}\ln\left[3\left(2\sqrt{3\lambda}\, t + \phi_0^3\right)^{2/3} - 4\right]$
$W(\phi) = \lambda\phi^4$ $V(\phi) = \lambda\phi^4 - \frac{8}{3}\lambda\phi^2$	$\Delta_N(\phi) = \frac{1}{6}\ln\left(3\phi^2 - 4\right), \quad \phi_0 = \pm\sqrt{5/3}$ $\Delta_N(t) = \frac{1}{6}\ln\left[3\phi_0^2 \exp\left(\pm 8\sqrt{\frac{\lambda}{3}}\, t\right) - 4\right]$

Given the expression for the *potential* by the *superpotential*

$$V = W - \frac{W'^2}{6W}, \tag{4.219}$$

we obtain

$$\Delta_N(\phi) = \int \frac{2W(W'^2 - W''W)}{W'(W'^2 - 2W''W + 6W^2)}\,d\phi. \tag{4.220}$$

Now, we define Δ_N as a function of the scalar field $\Delta_N = \Delta_N(\phi)$ from Eq. (4.220) and the time $\Delta_N = \Delta_N(t)$ by the dependence of the *scalar field* on time. The differences in the *e-folds number* for some inflationary models are presented in the Table (4.1). From the condition $\Delta_N(t = 0) = 0$ one can find the initial value of the *scalar field* ϕ_0.

Also, the differences Δ_N for the other models were considered in the paper [59].

Part III
Cosmological perturbations

Chapter 5

Cosmological perturbations

Cosmological perturbations are the catalysts for the evolution of large-scale structures in the universe. These patterns of perturbations in the metric are like fingerprints that unequivocally characterize a period of inflation. When matter fell in the troughs of these waves, it created density perturbations that collapsed gravitationally to form galaxies, clusters and superclusters of galaxies, with a spectrum that is also scale invariant. Such a type of spectrum was proposed in the early 1970s by Harrison [60] and Zel'dovich [61], to explain the distribution of galaxies and clusters of galaxies on very large scales in our observable *universe*.

The generation of initial perturbations has a quantum-mechanical nature. The quantum-mechanical generation of *cosmological perturbations* depends on only the existence of their quantum fluctuations at the initial point and interaction of perturbations with a variable gravitational field in a homogeneous isotropic *universe*. Fluctuations approached the scale of classical inhomogeneities, sufficient for formation in the further evolution of galaxies.

A strong alternating gravitational field during the very early *universe* plays the role of a "pump" field. It replaces the energy of zero quantum disturbances and increases them. The initial quantum state of each mode of perturbations is transformed as a result of the evolution of the quantum mechanical Schrödinger evolution into a state of a "frozen" vacuum.

For a chosen *cosmological model*, describing the *universe* during its inflationary stage, one can determine the *power spectrum* of density perturbations, the spectral indices of *scalar* and *tensor perturbations* and their ratio, as well as the ratio of the squared amplitudes of the tensor and scalar modes.

5.1 Inflationary parameters

Let us list the main parameters, which are defined in the models of cosmological *inflation*. The *state parameter* is defined as

$$w = \frac{p}{\rho} = \frac{\frac{1}{2}\dot{\phi}^2 - V}{\frac{1}{2}\dot{\phi}^2 + V} = -1 - \frac{2}{3}\frac{\dot{H}}{H^2} = -1 + \frac{4}{3}\left(\frac{H'_\phi}{H}\right)^2. \tag{5.1}$$

In *slow-roll approximation*, $\frac{1}{2}\dot{\phi}^2 \simeq 0$ and $\ddot{\phi} \simeq 0$. We have $w \simeq -1$, that corresponds to the accelerated expansion.

The *deceleration parameter*, which characterizes the rate of expansion of the *universe*, is defined as

$$q = -\frac{\ddot{a}a}{\dot{a}^2} = -1 - \frac{\dot{H}}{H^2} = -1 + 2\left(\frac{H'_\phi}{H}\right)^2. \tag{5.2}$$

In *slow-roll approximation*, $q \simeq -1$ also.

Further, we consider the parameters ϵ, δ, ξ which are important characteristics of the *inflation* stage and are directly related to the parameters of *cosmological perturbations* and with dynamical equations as well. In the literature they are called *slow-roll parameters*, but we will call them *exact inflationary parameters* to distinguish from the parameters determined from the *slow-roll approximation*.

Now, we will give a simple derivation of these parameters. For this purpose we write the dynamical equations in the following form

$$V(\phi) = 3H^2 - 2H'^2_\phi, \tag{5.3}$$

$$\dot{\phi} = -2H'_\phi. \tag{5.4}$$

The first parameter ϵ is obtained by dividing Eq. (5.3) by H^2, and as a result one has

$$\frac{V}{H^2} = 3 - 2\frac{H'^2_\phi}{H^2} = 3 - \epsilon, \quad \epsilon \equiv 2\frac{H'^2_\phi}{H^2}, \tag{5.5}$$

$$V = H^2(3 - \epsilon). \tag{5.6}$$

Further, we differentiate Eq. (5.3) with respect to ϕ

$$V'_\phi = 6H'_\phi H - 4H'_\phi H''_\phi \tag{5.7}$$

and divide by H^2

$$\frac{V'_\phi}{H^2} = 6\frac{H'_\phi}{H} - 4\frac{H'_\phi}{H}\frac{H''_\phi}{H}. \tag{5.8}$$

Then, we express, where it's possible, the terms in this equation by means of parameter ϵ:

$$\frac{V'_\phi}{H^2} = 6\sqrt{\frac{\epsilon}{2}} - 4\sqrt{\frac{\epsilon}{2}}\left(\frac{H''_\phi}{H}\right). \qquad (5.9)$$

From the last term we determine the second parameter δ:

$$\frac{V'_\phi}{H^2} = 6\sqrt{\frac{\epsilon}{2}} - 2\sqrt{\frac{\epsilon}{2}}\delta, \quad \delta \equiv 2\frac{H''_\phi}{H}. \qquad (5.10)$$

Substituting $\epsilon = 2\frac{H'^2_\phi}{H^2}$ into Eq. (5.10), we get

$$V'_\phi = 2H'_\phi H(3 - \delta). \qquad (5.11)$$

The relationship between the parameters ϵ and δ is obtained by differentiating Eq. (5.6) and equating it to (5.11). As a result we have

$$\delta = \epsilon + (\sqrt{2\epsilon})'_\phi. \qquad (5.12)$$

The third parameter is obtained similarly, differentiating (5.7) in ϕ and dividing by H^2 and expressing, where it is possible, the terms in this equation through

$$\frac{V''_\phi}{H^2} = -4\frac{H'''_\phi H'_\phi}{H^2} + 6\frac{H'^2_\phi}{H^2} + 6\frac{H''_\phi}{H} - 4\frac{H''^2_\phi}{H^2}$$

$$= -4\frac{H'''_\phi H'_\phi}{H^2} + 3(\epsilon + \delta) - \delta^2. \qquad (5.13)$$

The first term in this equation we denote as the third parameter ξ

$$\frac{V''_\phi}{H^2} = -\xi + 3(\epsilon + \delta) - \delta^2, \quad \text{where } \xi \equiv 4\frac{H'''_\phi H'_\phi}{H^2}. \qquad (5.14)$$

Thus, we obtain

$$V''_\phi = H^2[-\xi + 3(\epsilon + \delta) - \delta^2]. \qquad (5.15)$$

The relationship between the parameters ϵ, δ and ξ is obtained by differentiating Eq. (5.11) and equating it to (5.15), as a result

$$\xi = \delta'_\phi \sqrt{2\epsilon} + \epsilon\delta. \qquad (5.16)$$

Also, one can define ξ only through ϵ, substituting (5.12) in (5.16).

This procedure can be repeated to any order and the general formula for the *inflationary parameters* is

$$\epsilon_n = 2^n \frac{(H'_\phi)^{n-1}}{H^n} \frac{d^{(n+1)}H}{d\phi^{(n+1)}}, \quad \text{where } n \geq 1. \qquad (5.17)$$

Thus, the *exact inflationary parameters* ϵ, δ, ξ ($\epsilon \equiv \epsilon_H$, $\delta \equiv \delta_H$, $\xi \equiv \xi_H$) are

$$\epsilon \equiv 2\left(\frac{H'_\phi}{H}\right)^2 = -\frac{\dot{H}}{H^2}, \tag{5.18}$$

$$\delta \equiv 2\frac{H''_\phi}{H} = \epsilon - \frac{\dot{\epsilon}}{2H\epsilon} = -\frac{\ddot{H}}{2H\dot{H}}, \tag{5.19}$$

$$\xi \equiv 4\frac{H'_\phi H'''_\phi}{H^2} = \epsilon\delta - \frac{1}{H}\dot{\delta} = \frac{1}{2H^2}\frac{d}{dt}\left(\frac{\ddot{H}}{\dot{H}}\right). \tag{5.20}$$

The exact *e-folds number* or the number of increases in the size of the universe in *e*-times is defined as follows

$$N = \ln\frac{a_{end}}{a_i} = \int_{t_i}^{t_e} H dt = \int_{\phi_i}^{\phi_e} \frac{d\phi}{\sqrt{2\epsilon(\phi)}} = -\frac{1}{2}\int_{\phi_i}^{\phi_e} \frac{H d\phi}{H'_\phi}, \tag{5.21}$$

where t_i and t_e are the times of the beginning and the end of *inflation*.

The *inflationary parameters* in *slow-roll approximation* are

$$\epsilon \simeq \frac{1}{2}\left(\frac{V'_\phi}{V}\right)^2 \equiv \epsilon_V, \tag{5.22}$$

$$\delta \simeq \frac{V''_\phi}{V} - \frac{1}{2}\left(\frac{V'_\phi}{V}\right)^2 \equiv \eta_V - \epsilon_V, \tag{5.23}$$

$$\xi \simeq \frac{V'_\phi V'''_\phi}{V^2} - \frac{3}{2}\frac{V''_\phi}{V}\left(\frac{V'_\phi}{V}\right)^2 + \frac{3}{4}\left(\frac{V'_\phi}{V}\right)^4$$

$$\equiv \xi_V - 3\eta_V \epsilon_V + 3\epsilon_V^2. \tag{5.24}$$

The expressions (5.22)–(5.24) define *slow-roll parameters* ϵ_V, η_V, $\xi_V \ll 1$, and the *e-folds number* in *slow-roll approximation* is

$$N \simeq \int_{\phi_i}^{\phi_e} \frac{d\phi}{\sqrt{2\epsilon_V(\phi)}} = -\frac{1}{2}\int_{\phi_i}^{\phi_e} \frac{V d\phi}{V'_\phi}. \tag{5.25}$$

5.2 The perturbation of the scalar field and metric

Until now we have considered only the unperturbed FRW metric described by a *scale factor* $a(t)$ and a homogeneous *scalar field* $\phi(t)$.

Let us now consider the SCE's with perturbations. We choose the FRW metric in the form

$$ds^2 = a^2(\eta)[-d\eta^2 + \gamma_{ij} dx^i dx^j], \tag{5.26}$$

$$\phi = \phi(\eta), \tag{5.27}$$

where $\eta = \int dt/a(t)$ is the *conformal time*, in which the SCE's can be written as

$$3\mathcal{H}^2 = \frac{1}{2}\phi'^2 + a^2 V(\phi), \tag{5.28}$$

$$\mathcal{H}' - \mathcal{H}^2 = -\frac{1}{2}\phi'^2, \tag{5.29}$$

$$\phi'' + 2\mathcal{H}\phi' + a^2 V'(\phi) = 0, \tag{5.30}$$

where $\mathcal{H} = aH$, $\phi' = a\dot{\phi}$ and the prime denotes a derivative with respect to the *conformal time* η in this section.

During *inflation*, the quantum fluctuations of the *scalar field* will create perturbations of the metric. In the linear approximation, we write the perturbed metric taking into account *scalar* and *tensor perturbations* and field perturbations in linear perturbation theory [62]. We note that *inflation* cannot generate, to linear order, *vector perturbations*.

Under such circumstances, the metric takes the form

$$ds^2 = a^2(\eta)\Big[- (1 + 2A)d\eta^2 + 2B_{|i}dx^i d\eta$$

$$+ \Big\{(1 + 2\mathcal{R})\gamma_{ij} + 2E_{|ij} + 2h_{ij}\Big\}dx^i dx^j\Big], \tag{5.31}$$

$$\phi = \phi(\eta) + \delta\phi(\eta, x^i). \tag{5.32}$$

The indices $\{i, j\}$ denote the three-dimensional spatial coordinates with metric γ_{ij}, and $|i$ denotes the covariant derivative with respect to that metric. The gauge invariant *tensor perturbations* h_{ij} corresponds to a transverse traceless *gravitational wave*, $\nabla^i h_{ij} = h_i^i = 0$. The *scalar perturbations* $(A, B, \mathcal{R}, E)$ are gauge dependent functions of (η, x^i). Under a general coordinate (gauge) transformation [62], [29] one has

$$\tilde{\eta} = \eta + \xi^0(\eta, x^i), \tag{5.33}$$

$$\tilde{x}^i = x^i + \gamma^{ij}\xi_{|j}(\eta, x^i), \tag{5.34}$$

with arbitrary functions (ξ^0, ξ). The scalar and *tensor perturbations* transform, to linear order, as

$$\tilde{A} = A - \xi^{0'} - \mathcal{H}\xi^0, \quad \tilde{B} = B + \xi^0 - \xi', \tag{5.35}$$

$$\tilde{\mathcal{R}} = \mathcal{R} - \mathcal{H}\xi^0, \quad \tilde{E} = E - \xi, \tag{5.36}$$

$$\tilde{h}_{ij} = h_{ij}. \tag{5.37}$$

But it is possible to construct two *gauge-invariant gravitational potentials*
[62], [29]:

$$\Phi = A + (B - E')' + \mathcal{H}(B - E'), \tag{5.38}$$

$$\Psi = \mathcal{R} + \mathcal{H}(B - E'). \tag{5.39}$$

During *inflation*, the gauge-invariant equations for the perturbations on
comoving hypersurfaces of constant energy density are

$$\Phi'' + 3\mathcal{H}\Phi' + (\mathcal{H}' + 2\mathcal{H}^2)\Phi = \frac{1}{2}[\phi'\delta\phi' - a^2 V'(\phi)\delta\phi], \tag{5.40}$$

$$-\nabla^2\Phi + 3\mathcal{H}\Phi' + (\mathcal{H}' + 2\mathcal{H}^2)\Phi = -\frac{1}{2}[\phi'\delta\phi' + a^2 V'(\phi)\delta\phi], \tag{5.41}$$

$$\Phi' + \mathcal{H}\Phi = \frac{1}{2}\phi'\delta\phi, \tag{5.42}$$

$$\delta\phi'' + 2\mathcal{H}\delta\phi' - \nabla^2\delta\phi = 4\phi'\Phi' - 2a^2 V'(\phi)\Phi - a^2 V''(\phi)\delta\phi. \tag{5.43}$$

For simplification of this system of equations, new variables are usually
used [29]:

$$u \equiv a\delta\phi + z\Phi, \tag{5.44}$$

$$z \equiv a\frac{\phi'}{\mathcal{H}}. \tag{5.45}$$

Under this transformation, Eqs. (5.40)–(5.43) are reduced to three independent equations

$$u'' - \nabla^2 u - \frac{z''}{z}u = 0, \tag{5.46}$$

$$\nabla^2\Phi = \frac{\mathcal{H}}{2a^2}(zu' - z'u), \tag{5.47}$$

$$\left(a^2\frac{\Phi}{\mathcal{H}}\right)' = \frac{1}{2}zu. \tag{5.48}$$

From Eq. (5.46) one can find the function $u(z)$, which when substituted
into (5.48), can be integrated to give $\Phi(z)$, and together with $u(z)$ allow us
to obtain $\delta\phi$.

5.2.1   *Quantum origin of perturbations*

Now we consider the perturbations Φ and $\delta\phi$ as quantum fields. Note that
the perturbed action for the scalar mode u can be written as [29]:

$$\delta S = \frac{1}{2}\int d^3x\, d\eta\left[(u')^2 - (\nabla u)^2 + \frac{z''}{z}u^2\right]. \tag{5.49}$$

For quantization of the field u in the curved background defined by the metric (5.26), we can write the operator as

$$\hat{u}(\eta, \mathbf{x}) = \int d^3 \frac{\mathbf{k}}{(2\pi)^{3/2}} \left[u_k(\eta)\, \hat{a}_{\mathbf{k}}\, e^{i\mathbf{k}\cdot\mathbf{x}} + u_k^*(\eta)\, \hat{a}_{\mathbf{k}}^+\, e^{-i\mathbf{k}\cdot\mathbf{x}} \right], \tag{5.50}$$

where the creation and annihilation operators satisfy the common commutation relations which are defined by the vacuum condition

$$[\hat{a}_{\mathbf{k}}, \hat{a}_{\mathbf{k}'}^+] = \delta^3(\mathbf{k} - \mathbf{k}'), \tag{5.51}$$

$$\hat{a}_{\mathbf{k}}|0\rangle = 0. \tag{5.52}$$

However, we note that the *inflaton* can be written as a quantum field with its relations.

The equations of motion for each mode $u_k(\eta)$ are decoupled in linear perturbation theory

$$u_k'' + \left(k^2 - \frac{z''}{z} \right) u_k = 0. \tag{5.53}$$

The ratio z''/z acts like a time-dependent potential for the *Schrödinger like equation* (5.53). For finding *exact solutions* of the mode equation, we will use the *inflationary parameters* [63]:

$$\epsilon = 1 - \frac{\mathcal{H}'}{\mathcal{H}^2} = \frac{z^2}{2a^2}, \tag{5.54}$$

$$\delta = 1 - \frac{\phi''}{\mathcal{H}\phi'} = 1 + \epsilon - \frac{z'}{\mathcal{H}z}, \tag{5.55}$$

$$\xi = -\left(2 - \epsilon - 3\delta + \delta^2 - \frac{\phi'''}{\mathcal{H}^2\phi'} \right). \tag{5.56}$$

In terms of these parameters, the *conformal time* and the effective *potential* for the u_k mode can be written as

$$\eta = -\frac{1}{\mathcal{H}} + \int \epsilon \frac{da}{a\mathcal{H}}, \tag{5.57}$$

$$\frac{z''}{z} = \mathcal{H}^2 \left[(1 + \epsilon - \delta)(2 - \delta) + \mathcal{H}^{-1}(\epsilon' - \delta') \right]. \tag{5.58}$$

Considering the *inflationary parameters* (5.54) and (5.55) as constant to order ϵ^2 [64], we obtain

$$\epsilon' = 2\mathcal{H}\left(\epsilon^2 - \epsilon\delta \right) = \mathcal{O}(\epsilon^2), \tag{5.59}$$

$$\delta' = \mathcal{H}\left(\epsilon\delta - \xi \right) = \mathcal{O}(\epsilon^2), \tag{5.60}$$

$$\dot{\epsilon} = 2H\left(\epsilon^2 - \epsilon\delta\right) = \mathcal{O}(\epsilon^2), \tag{5.61}$$

$$\dot{\delta} = H\left(\epsilon\delta - \xi\right) = \mathcal{O}(\epsilon^2). \tag{5.62}$$

In that case, we can write

$$\eta = -\frac{1}{\mathcal{H}}\frac{1}{1-\epsilon}, \tag{5.63}$$

$$\frac{z''}{z} = \frac{1}{\eta^2}\left(\nu^2 - \frac{1}{4}\right), \quad \nu = 1 + \epsilon - \frac{\delta}{1-\epsilon} + \frac{1}{2}. \tag{5.64}$$

There is a characteristic scale given by the event horizon size or Hubble scale during *inflation*, H^{-1}. There will be modes u_k with physical wavelengths much smaller than this scale $k/a \gg H$, that are well within the de Sitter horizon, and therefore do not feel the curvature of space-time. Also, there are modes with wavelengths greater than the Hubble scale, $k/a \ll H$. In these two asymptotic regimes, the solutions can be written as

$$u_k = \frac{1}{\sqrt{2k}}e^{-ik\eta}, \quad k \gg aH, \tag{5.65}$$

$$u_k = C_1 z, \quad k \ll aH. \tag{5.66}$$

One can find, under the conditions (5.59)–(5.60), *exact solutions* of Eq. (5.53) with the *effective potential* given by (5.64) in the following form

$$u_k(\eta) = \frac{\sqrt{\pi}}{2}e^{i\left(\nu+\frac{1}{2}\right)\frac{\pi}{2}}(-\eta)^{1/2}H_\nu^{(1)}(-k\eta), \tag{5.67}$$

where $H_\nu^{(1)}(z)$ is the *Hankel function* of the first kind and ν is given by (5.64) in terms of the *slow-roll parameters*. In the limit $k\eta \to 0$, the solution becomes

$$|u_k| = \frac{2^{\nu-\frac{3}{2}}}{\sqrt{2k}}\frac{\Gamma(\nu)}{\Gamma\left(\frac{3}{2}\right)}(-k\eta)^{\frac{1}{2}-\nu} = \frac{C(\nu)}{\sqrt{2k}}\left(\frac{k}{aH}\right)^{\frac{1}{2}-\nu}, \tag{5.68}$$

$$C(\nu) = 2^{\nu-\frac{3}{2}}\frac{\Gamma(\nu)}{\Gamma\left(\frac{3}{2}\right)}(1-\epsilon)^{\nu-\frac{1}{2}}. \tag{5.69}$$

Let us now consider the *tensor perturbations* generated during *inflation*. The perturbed action for the tensor modes can be written as

$$\delta S = \frac{1}{2}\int d^3x\, d\eta\, \frac{a^2}{2}\left[(h_{ij}')^2 - (\nabla h_{ij})^2\right], \tag{5.70}$$

with the tensor field h_{ij} considered as a quantum field. The corresponding operator is

$$\hat{h}_{ij}(\eta, \mathbf{x}) = \int \frac{d^3 \mathbf{k}}{(2\pi)^{3/2}} \sum_{\lambda=1,2} \left[h_k(\eta)\, e_{ij}(\mathbf{k}, \lambda)\, \hat{a}_{\mathbf{k},\lambda}\, e^{i\mathbf{k}\cdot\mathbf{x}} + h.c. \right], \quad (5.71)$$

where $e_{ij}(\mathbf{k}, \lambda)$ are the two polarization tensors, satisfying symmetric, transverse and traceless conditions

$$e_{ij} = e_{ji}, \quad k^i e_{ij} = 0, \quad e_{ii} = 0, \quad (5.72)$$

$$e_{ij}(-\mathbf{k}, \lambda) = e_{ij}^*(\mathbf{k}, \lambda), \quad \sum_\lambda e_{ij}^*(\mathbf{k}, \lambda) e^{ij}(\mathbf{k}, \lambda) = const, \quad (5.73)$$

where the choice of the constant determines the normalization condition for the polarization tensors.

We can now redefine our gauge invariant tensor amplitude as

$$v_k(\eta) = \frac{a}{\sqrt{2\kappa}}\, h_k(\eta), \quad (5.74)$$

which satisfies the following evolution equation, decoupled for each mode $v_k(\eta)$ in linear perturbation theory:

$$v_k'' + \left(k^2 - \frac{a''}{a} \right) v_k = 0. \quad (5.75)$$

The ratio a''/a acts like a time-dependent *potential* for this *Schrödinger like equation*, analogous to the term z''/z for the *scalar metric perturbation*. For constant *inflationary parameters*, the *potential* is

$$\frac{a''}{a} = 2\mathcal{H}^2 \left(1 - \frac{\epsilon}{2} \right) = \frac{1}{\eta^2} \left(\mu^2 - \frac{1}{4} \right), \quad (5.76)$$

$$\mu = \frac{1}{1-\epsilon} + \frac{1}{2}. \quad (5.77)$$

One can solve Eq. (5.75) in the two asymptotic regimes,

$$v_k = \frac{1}{\sqrt{2k}}\, e^{-ik\eta}, \quad k \gg aH, \quad (5.78)$$

$$v_k = C\, a, \quad k \ll aH. \quad (5.79)$$

For constant *inflationary parameters*, one can find *exact solutions* of (5.75), with *effective potential* given by (5.76), that interpolate between the two asymptotic solutions. These are identical to Eq. (5.67) except for the substitution $\nu \to \mu$. In the limit $k\eta \to 0$, the solution becomes

$$|v_k| = \frac{C(\mu)}{\sqrt{2k}} \left(\frac{k}{aH} \right)^{\frac{1}{2} - \mu}, \quad (5.80)$$

with

$$C(\mu) = 2^{\mu - \frac{3}{2}} \frac{\Gamma(\mu)}{\Gamma\left(\frac{3}{2}\right)} (1 - \epsilon)^{\mu - \frac{1}{2}}. \tag{5.81}$$

Since the mode h_k becomes constant on superhorizon scales, one can evaluate the *tensor metric perturbation* when it reentered during the radiation or matter era directly in terms of its value during *inflation*.

5.2.2 Power spectra of scalar and tensor metric perturbations

Not only do we expect to measure the amplitude of the metric perturbations generated during inflation and responsible for the anisotropies in the CMB and density fluctuations, but we should also be able to measure its power spectrum, or two-point correlation function in Fourier space. Let us consider, firstly, the *scalar perturbations* $\mathcal{R}_k$. The two-point correlation can be given by [64]:

$$\langle 0|\mathcal{R}_k^* \mathcal{R}_{k'}|0\rangle = \frac{2\pi^2}{k^3} \mathcal{P}_\mathcal{R}(k) \delta^3(\mathbf{k} - \mathbf{k}') = \frac{|u_k|^2}{z^2} \delta^3(\mathbf{k} - \mathbf{k}'), \tag{5.82}$$

from $\mathcal{R}_k = \zeta_k = u_k/z$. Now, we write the *power spectrum* of the *scalar perturbations* without restrictions on the function $C(\nu)$ from Eqs. (5.68) and (5.82) as

$$\mathcal{P}_\mathcal{R}(k) = \frac{k^3}{2\pi^2} \frac{|u_k|^2}{z^2} = \frac{C^2(\nu)}{2\epsilon} \left(\frac{H}{2\pi}\right)^2 \left(\frac{k}{aH}\right)^{3 - 2\nu}. \tag{5.83}$$

We will consider the parameters of *cosmological perturbations* on the crossing of the *Hubble radius* $k = aH$.

On the crossing of the *Hubble radius* we have

$$\mathcal{P}_\mathcal{R}(k) = \frac{C^2(\nu)}{2\epsilon} \left(\frac{H}{2\pi}\right)^2 \Bigg|_{k=aH}. \tag{5.84}$$

For calculating a *spectral index* for *scalar perturbations*, we find, from condition $k = aH$, that

$$d\ln k = H dt + \frac{\dot{H}}{H} dt = H(1 - \epsilon) dt. \tag{5.85}$$

Also, we may extend it as

$$d\ln k = H(1 - \epsilon) dt = \frac{H}{\dot{\phi}}(1 - \epsilon) dt = \frac{H}{2H'_\phi}(\epsilon - 1) d\phi = \left(\frac{H'_\phi}{H} - \frac{H}{2H'_\phi}\right) d\phi. \tag{5.86}$$

And, from the definition of a *spectral index* for *scalar perturbations* or *scalar tilt*, we get

$$n_s - 1 \equiv \frac{d \ln \mathcal{P}_\mathcal{R}(k)}{d \ln k} = \frac{1}{H(1 - \epsilon)} \left[\frac{d \ln \mathcal{P}_\mathcal{R}(t)}{dt} \right]. \tag{5.87}$$

Thus, we obtain

$$n_s - 1 = -\frac{1}{H^2 \epsilon (1 - \epsilon)^3} \Big\{ 2\Psi(\nu) H \epsilon (\dot\epsilon \delta - \dot\delta \epsilon - 2\dot\epsilon + \dot\delta)$$

$$+ \dot\epsilon \epsilon H [2 \ln(1 - \epsilon)(\delta - 2\epsilon) + 2 \ln 2(\delta - 2) + 3\epsilon - 2\delta]$$

$$+ 2H\epsilon\dot\delta(1 - \epsilon) \ln[2(1 - \epsilon)] - 2\dot{H}\epsilon(1 - \epsilon)^2 + H\dot\epsilon \Big\}, \tag{5.88}$$

where $\Psi = \Psi(\nu)$ is the *digamma function*, which can be defined at a constant value of the parameter ν by the expression $\Psi(n) = H_{n-1} - \gamma$, where H_n is the harmonic number and $\gamma \approx 0.57722$ is *Euler constant*. Thus, $\Psi(\frac{3}{2}) = 2 - \gamma - 2 \ln 2$.

The *running* for *scalar perturbations* is denoted as

$$\frac{dn_s}{d \ln k} = \frac{1}{H(1 - \epsilon)} \left(\frac{dn_s}{dt} \right). \tag{5.89}$$

Let us consider now the *tensor or gravitational wave perturbations*

$$\sum_\lambda \langle 0 | h^*_{k,\lambda} h_{k',\lambda} | 0 \rangle = \frac{s}{a^2} |v_k|^2 \delta^3(\mathbf{k} - \mathbf{k}') \equiv \frac{2\pi^2}{k^3} P_g(k) \delta^3(\mathbf{k} - \mathbf{k}'). \tag{5.90}$$

The factor s depends on the choice of the normalization condition for polarization tensors of *gravitational waves* (5.73). In the literature, one can find that $s = 2$ [64,65] or $s = 8$ [63,66]. We will consider the first normalization with $s = 2$.

On the crossing of the *Hubble radius*, we have the *power spectrum* for *tensor perturbations*

$$P_g(k) = 2 \times C^2(\mu) \left(\frac{H}{2\pi} \right)^2_{k=aH}, \tag{5.91}$$

where Eqs. (5.74) and (5.80) were used. The factor 2 on the r.h.s. is due to the two polarizations.

The *spectral index* for *tensor perturbations* or *tensor tilt* is written as

$$n_g \equiv \frac{d \ln P_g(k)}{d \ln k} = \frac{1}{H(1 - \epsilon)} \left[\frac{d \ln P_g(t)}{dt} \right], \tag{5.92}$$

$$n_g = \frac{2}{H^2 \epsilon (1 - \epsilon)^3} \Big\{ H\dot\epsilon(\Psi(\mu) + \ln[2(1 - \epsilon)] - 1) + \dot{H}\epsilon(1 - \epsilon)^2 \Big\}. \tag{5.93}$$

The *running* for *tensor perturbations* is denoted as

$$\frac{dn_g}{d\ln k} = \frac{1}{H(1-\epsilon)}\left(\frac{dn_g}{dt}\right). \tag{5.94}$$

The *tensor-to-scalar ratio* is

$$r \equiv \frac{P_g}{P_{\mathcal{R}}} = 4\epsilon\frac{C^2(\mu)}{C^2(\nu)} = 4^{\frac{1-2\epsilon+\delta}{1-\epsilon}}\epsilon(1-\epsilon)^{\frac{2(\delta-\epsilon)}{1-\epsilon}}\frac{\Gamma^2(\mu)}{\Gamma^2(\nu)}. \tag{5.95}$$

Now we consider power-law *inflation* with *Hubble parameter* $H = m/t$, where m is a positive constant, $\epsilon = \delta = 1/m$ from (5.18)–(5.19), $\nu = \mu = \frac{1}{1-\epsilon} + \frac{1}{2}$ from (5.64) and (5.77), and

$$C(\nu) = C(\mu) = \left(2^{\frac{1}{1-\epsilon}-1}\right)\frac{\Gamma\left(\frac{1}{1-\epsilon}+\frac{1}{2}\right)}{\Gamma\left(\frac{3}{2}\right)}(1-\epsilon)^{\frac{1}{1-\epsilon}}, \tag{5.96}$$

from Eqs. (5.69) and (5.81).

Now, we obtain the parameters of the *cosmological perturbations* from Eqs. (5.84)–(5.95) as

$$P_{\mathcal{R}}(t) = \frac{m^3}{8\pi^2t^2}\left[\left(2^{\frac{1}{1-\epsilon}-1}\right)\frac{\Gamma\left(\frac{1}{1-\epsilon}+\frac{1}{2}\right)}{\Gamma\left(\frac{3}{2}\right)}(1-\epsilon)^{\frac{1}{1-\epsilon}}\right]^2, \tag{5.97}$$

$$P_g(t) = \frac{2m^2}{\pi^2t^2}\left[\left(2^{\frac{1}{1-\epsilon}-1}\right)\frac{\Gamma\left(\frac{1}{1-\epsilon}+\frac{1}{2}\right)}{\Gamma\left(\frac{3}{2}\right)}(1-\epsilon)^{\frac{1}{1-\epsilon}}\right]^2, \tag{5.98}$$

$$n_s - 1 = n_g = -\frac{2\epsilon}{1-\epsilon} = -\frac{1}{m-1}, \tag{5.99}$$

$$r = 4\epsilon = 4/m. \tag{5.100}$$

From the condition $k = aH = \dot{a}$, where $a = a_0t^m$, we obtain

$$t = \left(\frac{k}{a_0m}\right)^{\frac{1}{m-1}}. \tag{5.101}$$

Thus, the *power spectra* as a function of the wavenumbers are

$$P_{\mathcal{R}}(k) = \frac{m^3}{8\pi^2}\left[\left(2^{\frac{1}{1-\epsilon}-1}\right)\frac{\Gamma\left(\frac{1}{1-\epsilon}+\frac{1}{2}\right)}{\Gamma\left(\frac{3}{2}\right)}(1-\epsilon)^{\frac{1}{1-\epsilon}}\right]^2\left(\frac{k}{a_0m}\right)^{\frac{2}{1-m}}, \tag{5.102}$$

$$P_g(k) = \frac{2m^2}{\pi^2}\left[\left(2^{\frac{1}{1-\epsilon}-1}\right)\frac{\Gamma\left(\frac{1}{1-\epsilon}+\frac{1}{2}\right)}{\Gamma\left(\frac{3}{2}\right)}(1-\epsilon)^{\frac{1}{1-\epsilon}}\right]^2\left(\frac{k}{a_0m}\right)^{\frac{2}{1-m}}. \tag{5.103}$$

In the case of *slow-roll approximation*, $\epsilon \ll 1$ or $m \gg 1$ we have

$$\mathcal{P}_{\mathcal{R}}(k) \simeq \frac{m^3}{8\pi^2}\left(\frac{k}{a_0 m}\right)^{\frac{2}{1-m}}, \tag{5.104}$$

$$\mathcal{P}_{\mathcal{R}}(k) \simeq \frac{2m^2}{\pi^2}\left(\frac{k}{a_0 m}\right)^{\frac{2}{1-m}}, \tag{5.105}$$

$$n_s - 1 = n_g \simeq -2\epsilon = -2/m, \tag{5.106}$$

$$r = 4\epsilon = 4/m. \tag{5.107}$$

Consider the case when, on the crossing of the *Hubble radius*, the following conditions are fulfilled:

$$C(\nu) \simeq 1, \quad \nu \simeq 3/2, \tag{5.108}$$

$$C(\mu) \simeq 1, \quad \mu \simeq 3/2. \tag{5.109}$$

These conditions are realized in the case of *slow-roll approximation* $\epsilon, \delta \ll 1$ [63] and, also, with a special choice of parameters ϵ, δ. For example, in the case of $\delta = 2\epsilon + \Delta$, where the value of Δ is small compared with 2ϵ on the crossing of the *Hubble radius*, we have the condition (5.108).

Now, we note the parameters of *cosmological perturbations*[1] with conditions (5.108)–(5.109) from Eqs. (5.84)–(5.95)

$$\mathcal{P}_{\mathcal{R}}(k) = \frac{1}{2\epsilon}\left(\frac{H}{2\pi}\right)^2, \tag{5.110}$$

$$\mathcal{P}_g(k) = 2\left(\frac{H}{2\pi}\right)^2, \tag{5.111}$$

$$n_s - 1 = 2\left(\frac{\delta - 2\epsilon}{1 - \epsilon}\right) = -\frac{\left[2\epsilon + \frac{1}{H}\left(\frac{\dot{\epsilon}}{\epsilon}\right)\right]}{1 - \epsilon} = -\frac{\left[2\epsilon + \frac{d\ln\epsilon}{dN}\right]}{1 - \epsilon}, \tag{5.112}$$

$$n_g = -\frac{2\epsilon}{1 - \epsilon}, \tag{5.113}$$

$$\frac{dn_s}{d\ln k} = -\frac{2(2\delta^2\epsilon - \delta\epsilon^2 - 5\delta\epsilon + 4\epsilon^2 - \epsilon\xi + \xi)}{(1 - \epsilon)^3}, \tag{5.114}$$

$$\frac{dn_g}{d\ln k} = -\frac{4(\epsilon^2 - \epsilon\delta)}{(1 - \epsilon)^3}, \tag{5.115}$$

$$\frac{n_s - 1}{n_g} = 2 - \frac{\delta}{\epsilon}, \quad r = 4\epsilon. \tag{5.116}$$

---

[1]For $s = 8$ [63, 66] one has $\mathcal{P}_g(k) = 8(H/2\pi)^2$ and $r = 16\epsilon$.

The *spectral index* of *scalar perturbations* can be both negative and positive, but the *spectral index* of *tensor perturbations* is only negative.

Let us remember that in the *slow-roll approximation* we have

$$\mathcal{P}_{\mathcal{R}}(k) = \frac{1}{2\epsilon}\left(\frac{H}{2\pi}\right)^2 \simeq \frac{1}{2\epsilon_V}\left(\frac{H}{2\pi}\right)^2, \tag{5.117}$$

$$\mathcal{P}_g(k) = 2\left(\frac{H}{2\pi}\right)^2, \tag{5.118}$$

$$n_s - 1 \simeq 2(\delta - 2\epsilon) \simeq 2\eta_V - 6\epsilon_V, \tag{5.119}$$

$$n_g \simeq -2\epsilon \simeq -2\epsilon_V, \tag{5.120}$$

$$\frac{dn_s}{d\ln k} \simeq -(2\xi + 8\epsilon^2 - 10\epsilon\delta) \simeq -(2\xi_V + 24\epsilon_V^2 - 16\eta_V\epsilon_V), \tag{5.121}$$

$$\frac{dn_g}{d\ln k} \simeq -4(\epsilon^2 - \epsilon\delta) \simeq -(8\epsilon_V^2 - 4\eta_V\epsilon_V), \tag{5.122}$$

$$\frac{n_s - 1}{n_g} = 2 - \frac{\delta}{\epsilon} \simeq 3 - \frac{\eta_V}{\epsilon_V}, \quad r = 4\epsilon \simeq 4\epsilon_V. \tag{5.123}$$

The running coincides modulo with that presented in [63], [66].

Since the *slow-roll approximation* can be used at different stages of calculating the parameters of *cosmological perturbations*, we will call the parameters of *cosmological perturbations* obtained from Eqs. (5.84)–(5.95) as *exact*, from Eqs. (5.110)–(5.116) as *refined* and from Eqs. (5.117)–(5.123) calculated by means of ϵ, δ, ξ as *approximate*. For verification of the parameters of *cosmological perturbations*, we will use their refined expressions and the experimental data [67]

$$10^9\mathcal{P}_S = 2.142 \pm 0.049, \mathcal{P}_T = r\mathcal{P}_S,$$

$$n_S = 0.9667 \pm 0.0040,$$

$$r < 0.112.$$

One can determine the amplitude of the tensor mode and estimate its contribution to the *CMB anisotropy* on the basis of the *tensor-to-scalar ratio r*, and the observed contribution of the scalar mode for the considered model of cosmological *inflation*.

5.3   Calculation of cosmological parameters

Now, we will calculate the basic parameters of *cosmological perturbations* for some models of *inflation* from the *scale factor* by means of Eqs. (5.110)–(5.116).

5.3.1 *Power-law inflation*

Consider, first of all, power-law *inflation*. The *scale factor* in this model is defined as follows:

$$a(t) = a_0 t^m, \quad a_0 = const. \tag{5.124}$$

For the *spectral index* of *scalar* and *tensor perturbations*, we get

$$n_S - 1 = \frac{2}{1-m} = n_G. \tag{5.125}$$

Assuming $m > 1$ and $|n_S - 1| < 0.2$, we get a reasonable limit for the degree of expansion: $m > 9$. The *tensor-to-scalar ratio*

$$r = \frac{2}{3-m} \tag{5.126}$$

is critically associated with $m = 3$. If $m = 3$, we have $r \to \infty$. For $1 \le m < 3$, the *tensor-to-scalar ratio* is $1 \le r < \infty$, when $m > 3$, we obtain $1 \le r < \infty$. In this case, when $m \ge 9$, we have $r < 0.333$ and as m grows, the *tensor-to-scalar ratio* becomes smaller.

5.3.2 *de Sitter solutions*

The *de Sitter solution* describes singular and non-singular *universe* with *scale factors*

$$a(t) = a_0 \sinh(h_* t), \quad a_0 = constant, \tag{5.127}$$

and

$$a(t) = a_0 \cosh(h_* t), \tag{5.128}$$

respectively. *Power spectra* and their *tilts* are calculated by general formulas (5.110)–(5.123), and have the following expressions:

– for a singular *universe*, we have

$$n_S - 1 = \frac{2[2 + \cosh^2(h_* t)]}{\sinh^2(h_* t)}, \tag{5.129}$$

$$n_G = -\frac{2}{\sinh^2(h_* t)}, \tag{5.130}$$

$$\mathcal{P}_\mathcal{R} = \frac{\coth^4(h_* t)}{8\pi^2 h_* [1 - \coth^2(h_* t)]}, \tag{5.131}$$

$$\mathcal{P}_g = \frac{\coth^2(h_* t)}{2\pi^2}, \quad r = 4\frac{h_* [1 - \coth^2(h_* t)]}{\coth^2(h_* t)}. \tag{5.132}$$

- in case of a non-singular *universe*, we get

$$n_S - 1 = \frac{2[2 - \sinh^2(h_*t)]}{\cosh^2(h_*t)}, \tag{5.133}$$

$$n_G = \frac{2}{\cosh^2(h_*t)}, \tag{5.134}$$

$$\mathcal{P}_{\mathcal{R}} = \frac{\tanh^4(h_*t)}{8\pi^2 h_*[1 - \tanh^2(h_*t)]}, \tag{5.135}$$

$$\mathcal{P}_G = \frac{\tanh^2(h_*t)}{2\pi^2}, \quad r = 4\frac{h_*[1 - \tanh^2(h_*t)]}{\tanh^2(h_*t)}. \tag{5.136}$$

The *scalar field* is imaginary in this case.

5.3.3 *Generalized exponential inflation*

Consider the *scale factor* defined as follows:

$$a = a_0 e^{(Ae^{\lambda t} + Bt)}, \tag{5.137}$$

where A, B, λ are constants.

The *Hubble parameter* is

$$H(t) = A\lambda e^{\lambda t} + B. \tag{5.138}$$

We calculate the *spectral indices* of *scalar* and *tensor perturbations*, as well as define the *power spectra*:

$$n_S - 1 = \frac{4A\lambda e^{\lambda t} - \lambda(A\lambda e^{\lambda t} + B)}{A\lambda^2 e^{\lambda t} + (A\lambda e^{\lambda t} + B)^2}, \tag{5.139}$$

$$n_G = \frac{2A\lambda^2 e^{\lambda t}}{A\lambda^2 e^{\lambda t} + (A\lambda e^{\lambda t} + B)^2}, \tag{5.140}$$

$$\mathcal{P}_{\mathcal{R}} = \frac{(A\lambda e^{\lambda t} + B)^4}{8\pi^2 A\lambda^2 e^{\lambda t}}, \quad \mathcal{P}_G = \frac{(A\lambda e^{\lambda t} + B)^2}{2\pi}, \tag{5.141}$$

$$r = 4\frac{A\lambda^2 e^{\lambda t}}{(A\lambda e^{\lambda t} + B)^2}. \tag{5.142}$$

5.3.4 *Exponential-power-law inflation*

Evolution of the *scale factor* in this model is defined as follows:

$$a(t) = a_0 t^n \exp(h_*t), \quad a_0 = const. \tag{5.143}$$

The parameters of *cosmological perturbations* in this case are written as

$$n_S - 1 = \frac{2(h_* t - n)}{(h_* t + n)^2 - n}, \quad n_G = \frac{2n}{n - (h_* t + n)^2}, \quad (5.144)$$

$$\mathcal{P}_{\mathcal{R}} = \frac{t^4}{8\pi^2 (h_* t + n)^2 [1 - t(h_* t + n)]}, \quad (5.145)$$

$$\mathcal{P}_{\mathcal{G}} = \frac{t^2}{2\pi^2 (h_* t + n)^2}. \quad (5.146)$$

We calculate the parameters of *cosmological perturbations* for the evolution of the *scale factor*

$$a = a_0 e^{\lambda(t - t_0)^\mu}, \quad \lambda, \ \mu\text{-constants.} \quad (5.147)$$

By calculations similar to the above, we get

$$n_S - 1 = \frac{3\mu + 1}{\mu(t - t_0)^\mu + \mu - 1}, \quad n_G = \frac{2}{1 + \left(\frac{\mu}{\mu - 1}\right)(t - t_0)^\mu}, \quad (5.148)$$

$$\mathcal{P}_{\mathcal{R}} = \frac{\lambda^3 \mu^3}{8\pi^2} \frac{(t - t_0)^{3\mu - 2}}{\mu - 1}, \quad \mathcal{P}_{\mathcal{G}} = \frac{\mu^2 \lambda^2}{2}(t - t_0)^{2\mu - 2}, \quad (5.149)$$

$$r = \frac{4(\mu - 1)}{\mu\lambda(t - t_0)^\mu}. \quad (5.150)$$

5.4 Cosmological parameters in conformally flat space

Based on the results obtained in Sec. 2.3, we write the exact values of the cosmological parameters at the crossing of the *Hubble radius* ($k = aH = \mathcal{H}$) by the *conformal factor*, given that

$$d\eta = \frac{dt}{a}, \mathcal{H} = \frac{a'}{a}, H = \frac{\mathcal{H}}{a}, \dot{H} = \frac{\mathcal{H}'}{a^2} - \frac{\mathcal{H}^2}{a^2}. \quad (5.151)$$

Also, let us remember, that we introduced the function $A(\eta) = a^2(\eta)$ as in (5.52).

As the result, from Eqs. (5.110)–(5.116) we get:

- *power spectra* of the *scalar* and *tensor perturbations*

$$\mathcal{P}_{\mathcal{R}}(k) = -\frac{\mathcal{H}^4 a^2}{8(\mathcal{H}' - \mathcal{H}^2)} = \frac{A'^4}{32\pi^2 A^3 (3A'^2 - 2A''A)} \quad (5.152)$$

$$\mathcal{P}_{\mathcal{G}}(k) = \frac{\mathcal{H}^2}{2\pi^2 a^2} = \frac{A'^2}{8\pi^2 A^3} \quad (5.153)$$

- *spectral indices* of *scalar* and *tensor perturbations*

$$n_S(k) - 1 = \frac{a^2}{\mathcal{H}'}\left[4\left(\frac{\mathcal{H}' - \mathcal{H}^2}{a^2}\right) - \frac{\mathcal{H}}{\mathcal{H}' - \mathcal{H}^2}\left(\frac{\mathcal{H}' - \mathcal{H}^2}{a^2}\right)'\right]$$

$$= \frac{9A'^4 + 8A''^2 A^2 - 14A'' A'^2 A - 2A''' A' A^2}{(2A'' A - 3A'^2)(A'' A - A'^2)} \tag{5.154}$$

$$n_{\mathcal{G}}(k) = \frac{2(\mathcal{H}' - \mathcal{H}^2)}{\mathcal{H}'} = \frac{2A'' A - 3A'^2}{A'' A - A'^2} \tag{5.155}$$

- *tensor-to-scalar ratio*

$$r = -4\left(\frac{\mathcal{H}' - \mathcal{H}^2}{\mathcal{H}^2}\right) = \frac{12A'^2 - 8A'' A}{A'^2} \tag{5.156}$$

When we calculate the cosmological parameters, the *conformal time* is the time of crossing of the *Hubble radius*.

We calculated the basic cosmological parameters for the selected *conformal factor* $A(\eta) = A_0 e^{\beta(\eta)}$ (Subsec. 2.3.1) at the crossing of the *Hubble radius* by the formulas (5.152)–(5.156)

$$\mathcal{P}_{\mathcal{R}}(k) = \frac{\beta'^4}{32\pi^2 A_0 e^\beta(\beta'^2 - 2\beta'')}, \tag{5.157}$$

$$\mathcal{P}_{\mathcal{G}}(k) = \frac{\beta'^2}{8\pi^2 A_0 e^\beta}, \tag{5.158}$$

$$n_S(k) - 1 = \frac{\beta'^4 + 8\beta''^2 - 4\beta'' \beta'^2 - 2\beta''' \beta'}{\beta''(2\beta'' - \beta'^2)}, \tag{5.159}$$

$$n_{\mathcal{G}}(k) = \frac{2\beta'' - \beta'^2}{\beta''}, \tag{5.160}$$

$$r = \frac{4\beta'^2 - 8\beta''}{\beta'^2}. \tag{5.161}$$

The *conformal time* of the crossing of the *Hubble radius* η_H for perturbations of a given wavelength (wave number) is determined from the condition $k = \mathcal{H}$, which is equivalent to the condition $2k = \beta'(\eta)|_{\eta=\eta_H}$ for the given *conformal factor*.

5.4.1   *Conformal power-law expansion*

We calculate the cosmological parameters for the model of Subsec. 2.3.2

$$\mathcal{P}_{\mathcal{R}}(k) = \frac{m^3(\alpha\eta_H)^{-m}}{32\pi^2 A_0 \eta_H^2(m + 2)}, \tag{5.162}$$

$$P_{\mathcal{G}}(k) = \frac{m^2 (\alpha \eta_H)^{-m}}{8\pi^2 A_0 \eta_H^2}, \tag{5.163}$$

$$n_S(k) - 1 = m + 2, \tag{5.164}$$

$$n_{\mathcal{G}}(k) = m + 2, \tag{5.165}$$

$$r = \frac{4(m+2)}{m}, \tag{5.166}$$

where the *conformal time* of crossing of the *Hubble radius* is $\eta_H = m/2k$.

To comply with the observed cosmological parameters, the obtained parameters should be satisfied, *viz.*, $-2.04 < m < -2$ [67].

5.4.2 *Generalized conformal exponential expansion*

The parameters of *cosmological perturbations* for the model of Subsec. 2.3.3 are

$$P_{\mathcal{R}}(k) = \frac{(c_1 \eta_H + c_2)^4}{32\pi^2 A_0 (c_1^2 \eta_H^2 + 2c_1^2 c_2 \eta_H + c_2^2 - 2c_1)} e^{-\left(\frac{c_1}{2}\eta_H^2 + c_2 \eta_H + c_3\right)}, \tag{5.167}$$

$$P_{\mathcal{G}}(k) = \frac{(c_1 \eta_H + c_2)^2}{8\pi^2 A_0} e^{-\left(\frac{c_1}{2}\eta_H^2 + c_2 \eta_H + c_3\right)}, \tag{5.168}$$

$$n_S(k) - 1 = \frac{4c_1^3 \eta_H^2 + 8c_1^2 c_2 \eta_H + 4c_1 c_2^2}{c_1(c_1^2 \eta_H^2 + 2c_1^2 c_2 \eta_H + c_2^2 - 2c_1)}$$
$$- \frac{4c_1^3 c_2 \eta_H^3 + 6c_1^2 c_2^2 \eta_H^2 + 4c_1 c_2^3 \eta_H + c_1^4 \eta_H^4 + c_2^4 + 8c_1^2}{c_1(c_1^2 \eta_H^2 + 2c_1^2 c_2 \eta_H + c_2^2 - 2c_1)}, \tag{5.169}$$

$$n_{\mathcal{G}}(k) = -c_1 \eta_H^2 - 2c_2 \eta_H - \frac{c_2^2}{c_1} + 2, \tag{5.170}$$

$$r = 4\frac{c_1^2 \eta_H^2 + 2c_1^2 c_2 \eta_H + c_2^2 - 2c_1}{(c_1 \eta_H + c_2)^2}, \tag{5.171}$$

where $\eta_H = \frac{c_2 - 2k}{c_1}$. For the function $\beta(\eta) = \frac{b_1}{12}\eta^4 + b_3$, we obtain the cosmological parameters

$$P_{\mathcal{R}}(k) = \frac{b_1^3 \eta_H^{10}}{288\pi^2 A_0 (b_1 \eta_H^4 - 18)} e^{-\left(\frac{b_1}{12}\eta_H^4 + b_3\right)}, \tag{5.172}$$

$$P_{\mathcal{G}}(k) = \frac{b_1^2 \eta_H^6}{72\pi^2 A_0} e^{-\left(\frac{b_1}{12}\eta_H^4 + b_3\right)}, \tag{5.173}$$

$$n_S(k) - 1 = \frac{540 - 38 b_1 \eta_H^4 + b_1^2 \eta_H^8}{9(b_1 \eta_H^4 - 18)}, \tag{5.174}$$

$$n_g(k) = 2 - \frac{b_1}{9} \eta_H^4, \tag{5.175}$$

$$r = 4 \frac{b_1 \eta_H^4 - 18}{b_1 \eta_H^4}, \tag{5.176}$$

where $\eta_H = (6k/b_1)^{1/3}$.

5.5 Post-inflationary evolution of cosmological perturbations

The *inflationary* stage is completed by *scalar field* decay and particle production, followed by nucleosynthesis and further evolution according to the standard scenario. *Cosmological perturbations* of different wavelengths (with different wavenumbers k) become classical quantities during several e-folds after the instant of coming outside the horizon. This time is designated as t_*. Having crossed the horizon, the *cosmological perturbations* remain "frozen" in the gravitational background and do not change their magnitude in the comoving reference frame. The magnitude of perturbations evolves in a known way in the radiation-dominated epoch at the time of entering under the horizon, and we denote this time as t_{pr}.

The theory of *cosmological perturbations* is considered to be applicable at the initial epoch which begins before the cosmological scale of interest enters the horizon. This initial epoch begins much later than nucleosynthesis, and therefore, the matter constituents of the *universe*, except nonbaryonic *dark matter*, are known. It has been established theoretically how perturbations of all components of the *universe* evolve after the initial epoch evolve, if the energy density of each component at that time is known.

The initial perturbation spectrum can be constructed using the *transfer function* from vacuum fluctuations at the instant t_*. It is known how to calculate this function; in doing so, the perturbation of each component is calculated from the curvature perturbation $\mathcal{R}_{\mathbf{k}}$ with the aid of the *transfer function*:

$$g_{\mathbf{k}}(t) = T_g(t, k) \mathcal{R}_{\mathbf{k}},$$

where $\mathcal{R}_{\mathbf{k}}(t)$ is determined at the time t_*:

$$\mathcal{R}_{\mathbf{k}} = -\left[\frac{H}{\dot{\phi}} \delta\phi_k\right]_{t=t_*}.$$

Our improvement is associated with calculating the curvature perturbation on the basis of *exact solutions* to the *Einstein equations* (in the
zero-order approximation) without using a slow-roll regime in this situation. Since we have obtained exact expressions for the cosmological parameters at an exit outside the horizon, we need, for confrontation with the
observational data, to carry out a re-calculation of the cosmological parameters at the present epoch. To do that, let us consider the post-inflationary
evolution of the *cosmological perturbations* and find correction to the cosmological parameters.

One can now compute Φ and $\delta\phi$ from the solution (5.66), for $k \ll aH$.
Substituting into Eq. (5.48), we obtain

$$\Phi = C_1 \left(1 - \frac{\mathcal{H}}{a^2} \int a^2 d\eta\right) + C_2 \frac{\mathcal{H}}{a^2}, \tag{5.177}$$

$$\delta\phi = \frac{C_1}{a^2} \int a^2 d\eta - \frac{C_2}{a^2}. \tag{5.178}$$

The term proportional to C_1 corresponds to the growing modes, and the
one proportional to C_2 corresponds to the decaying modes, which can be
ignored. In the case of adiabatic perturbations we get

$$\zeta \equiv \Phi + \frac{1}{\epsilon\mathcal{H}}\left(\Phi' + \mathcal{H}\Phi\right) = \frac{u}{z}, \tag{5.179}$$

which is constant for $k \ll aH$. This quantity ζ is identical, for superhorizon modes, to the gauge invariant curvature metric perturbation $\mathcal{R}_c$ on
comoving hypersurfaces [62], [68], [66]:

$$\zeta = \mathcal{R}_c + \frac{1}{\epsilon\mathcal{H}^2} \nabla^2 \Phi. \tag{5.180}$$

Using Eq. (5.47) we can write the evolution equation for $\zeta = \frac{u}{z}$ as
$\zeta' = \frac{1}{\epsilon\mathcal{H}} \nabla^2 \Phi$, which confirms that ζ is constant for adiabatic superhorizon
modes, $k \ll aH$. Therefore, one can evaluate the gravitational potential Φ_k
when the perturbation reenters the horizon during the radiation or matter
eras in terms of the curvature perturbation $\mathcal{R}_k$ when it left the Hubble scale
during inflation:

$$\Phi_k = \left(1 - \frac{\mathcal{H}}{a^2} \int a^2 d\eta\right) \mathcal{R}_k = \frac{3 + 3w}{5 + 3w} \mathcal{R}_k = \begin{cases} \dfrac{2}{3} \mathcal{R}_k & \text{radiation era,} \\[2mm] \dfrac{3}{5} \mathcal{R}_k & \text{matter era.} \end{cases}$$

We conclude from these relations that the evolution of perturbations at superhorizon scales is reduced to *rescaling* their amplitudes.

Thus, the *power spectrum* of the gravitational perturbations during the matter-dominated stage is determined through the *power spectrum* of curvature perturbations:

$$\mathcal{P}_\Phi(MD) = \frac{9}{25}\mathcal{P}_\mathcal{R}. \tag{5.181}$$

This allows us to determine the density contrast and the *power spectrum* of *scalar perturbations*

$$\delta_\mathbf{k} = \frac{2}{3}\left(\frac{k}{aH}\right)^2 T_g(t,k)\Phi_\mathbf{k}, \tag{5.182}$$

$$\mathcal{P}_\delta(k,t) = \frac{4}{25}\left(\frac{k}{aH}\right)^4 T_g(t,k)^2\mathcal{P}_\mathcal{R},$$

which can be compared with the observational data.

In the case of arbitrary wavelengths of the perturbations, the *transfer function* can be calculated numerically. The values of the *transfer function* T_g have been tabulated for many *cosmological models* [69], [70].

Thus our improvement introduced during the after-inflationary stage leads finally to improved relations containing observational data.

Part IV

Friedmann vs Abel equations: A connection

Chapter 6

The Abel equation and how to derive the unknown superpotentials

6.1 Setting up the stage

In this chapter we are going to take a next big step in our exploration of the *superpotential* W and of its role in the cosmology of a flat universe. We have already seen how short and simple the calculations become once we learn the exact shape of the *superpotential*; we have seen virtually every cosmological parameter of import being easily derived from $W(\varphi)$ alone. However, an observant reader has no doubt noticed that there remains one last physically important quantity whose derivation we have neglected to discuss so far: the *superpotential* itself! On the one hand, we already know it to be identical to the density ρ of a scalar field φ; so on a first glance its derivation must be straightforward: just plug in the known functional $\rho(\varphi)$ and be done with it! Unfortunately for us, this blissful prospect gets ruined by a simple fact that it is this very dependency that we usually do not possess. Instead, it is much more common to have a model with a known potential of a scalar field $V(\varphi)$, whose exact shape is predicted by the high energy physics. So, what to do? The knowledge we have gained so far is virtually useless; we can derive V from a given W, but can not go the other way around.

Thus, we arrive at an apparent dead end. We can't use our formulas since we don't know the *superpotential*, and we can not derive it as we do not possess a simple concise way to produce the *superpotential* from the potential V. So, we have no choice but to once again revisit the Friedman equations to see what we might have missed.

As we remember, the *Einstein-Friedman equations* has the form of

$$H^2 = \frac{8\pi}{3M_P^2}\left(\frac{1}{2}\dot{\varphi}^2 + V(\varphi)\right) - \frac{k}{a^2},$$

$$\ddot{\varphi} = -3H\dot{\varphi} - dV(\varphi)/d\varphi, \tag{6.1}$$

where $k = 0, +1, -1$ correspond to either a spatially-flat, closed or open Friedman universe, $a = a(t)$ is a scale factor, H is the Hubble parameter, φ is a scalar field, and V is its potential.[1]

As usual, we define the superpotential W as (see [53, 54, 71]):

$$W = \frac{1}{2}\dot{\varphi}^2 + V(\varphi). \tag{6.2}$$

A couple of observations are in order. First of all, we will assume the *superpotential* W to be nonnegative for all φ, thus ensuring that the *weak energy condition* $\rho \geq 0$ holds everywhere. The (6.2) then implies that for any *negatively defined* V, the kinetic term $\dot{\varphi}^2$ must be strictly positive. However, choosing a *positively defined* potential $V > 0$ alleviates this requirement, thus producing two distinct possibilities: either both $V > 0$ and $\dot{\varphi}^2 > 0$, or $V > 0$ while $\dot{\varphi}^2 < 0$. The former case provides a classical cosmological dynamics, but the latter leads to something quite different, as it describes the state of a universe that can be appropriately called a *phantom zone*, that is, the universe whose dynamics is propagated by the *phantom fields* with a negative pressure p ([72], [73], [74]). As we shall soon see, all of these possibilities will manifest itself in our calculation.

Let us assume that $k = 0$ (i.e. that the universe is spatially flat). The system (6.1) can be easily transformed into:

$$\frac{dW}{d\varphi} = -3H\dot{\varphi},$$

$$H = \pm\frac{1}{M_P}\sqrt{\frac{8\pi}{3}W}, \tag{6.3}$$

$$\frac{1}{2}\dot{\varphi}^2 + V(\varphi) = W(\varphi).$$

It immediately follows from (6.3) that the quantities $\dot{\varphi}$, H and V can all be rewritten in terms of the *superpotential* $W(\varphi)$ and its first

---

[1]The system (6.1) is written in the "natural" unit system in which $\hbar = c = 1$ and $G = M_P^{-2}$.

derivative $W'(\varphi)$:

$$H = \pm\alpha\sqrt{W},$$

$$\dot{\varphi} = \mp\frac{1}{3\alpha}\frac{W'}{\sqrt{W}},\tag{6.4}$$

$$V = W - \frac{1}{2}\left(\frac{1}{3\alpha}\frac{W'}{\sqrt{W}}\right)^2,$$

where we have introduced a positive constant α which satisfies the condition

$$\alpha^2 = \frac{8\pi}{3M_p^2}.$$

As we can see from (6.4), somewhat contrary to our assumptions, the cosmological quantities do not, in fact, depend on W itself: it turns out that they instead depend on a square root of W (and on its first derivative w.r.t. φ). Indeed, we can define two constant-sign axillary functions $R_\pm(\varphi)$ as:

$$R_\pm(\varphi) = \pm\sqrt{W(\varphi)},\tag{6.5}$$

so that the system (6.4) can be appropriately rewritten as:

$$H = \alpha R_\pm,$$

$$\dot{\varphi} = -\frac{2}{3\alpha}R'_\pm,\tag{6.6}$$

$$V = (R_\pm)^2 - 2\left(\frac{1}{3\alpha}R'_\pm\right)^2.$$

We are interested in the last one of those equations. It is, unfortunately, nonlinear: a difficult one to study and solve for practically any V, with a notable exception of a constant potential $V = \text{const}$: for example, the zero-valued potential $V \equiv 0$ renders the system (6.6) easily solvable:

$$R = \frac{H_0}{\alpha}e^{\pm\frac{3\alpha}{\sqrt{2}}(\varphi-\varphi_0)}$$

$$\varphi = \varphi_0 \mp \frac{\sqrt{2}}{3\alpha}\ln\left(1 + 3H_0(t - t_0)\right)$$

$$H = \frac{H_0}{1 + 3H_0(t - t_0)},$$

where $H_0 = H(t_0)$ and $\varphi_0 = \varphi(t_0)$. The general case of $V = \text{const} \neq 0$ can be resolved similarly. Unfortunately, any kind of φ dependence of V turns the system (6.6) into a one that can not be cracked as easily. Therefore, if we are to study the cosmological dynamics governed by such physically meaningful potentials as $V = m^2\varphi^2/2$ or $V = \lambda\varphi^4$, we will first have to transform the system (6.6) into something a little more manageable.

Before we do that, however, let us point out that by switching from the variable φ to a new variable x:

$$\varphi = \beta x, \qquad \beta = \text{const}, \tag{6.7}$$

we can transform (6.6) into:

$$H = \alpha R_{\pm},$$

$$\dot{x} = -\frac{2}{3\alpha\beta} R'_{\pm}(x),$$

$$v(x) = (R_{\pm}(x))^2 - 2\left(\frac{1}{3\alpha\beta} R'_{\pm}(x)\right)^2, \tag{6.8}$$

where $v(x) = V(\beta x)$, and β will play a role of a free parameter. As we will see in the next section, a proper choice of this parameter will mean nothing short of determining an exact dynamics of the universe.

6.2   The curious case of the Abel equation

We have discussed in the previous section that the cosmological dynamics of the universe strongly depend on the sign of the potential $v(x)$. This, combined with the weak energy condition $W \geq 0$, motivates us to consider the following relationship:

$$W(x) = \begin{cases} v(x)\, \Omega^2(x), & v \geq 0 \\ -v(x)\, \Omega^2(x), & v < 0 \end{cases} \tag{6.9}$$

where $\Omega(x)$ is some (yet) undefined real-valued function. Under this assumption, the axillary function R turns into

$$R(x) = \begin{cases} \sqrt{v(x)}\, \Omega(x), & v \geq 0 \\ \sqrt{-v(x)}\, \Omega(x), & v < 0 \end{cases} \tag{6.10}$$

where we have augmented the $\pm$ sign into Ω.

We should not be deceived by the simplicity of (6.10); after all, the function Ω must implicitly depend on v! In order to figure out how, let us look at the last equation of (6.8). Substituting (6.10) into it and assuming that $v \neq 0$ (we have already discussed the case $v \equiv 0$ in the previous section), this single equation morphs into the following system:

$$\begin{cases} 1 = \Omega^2 - \dfrac{1}{2(3\alpha\beta)^2} \left(2\Omega' + \dfrac{v'}{v}\Omega\right)^2, & v > 0 \\[4mm] 1 = -\Omega^2 + \dfrac{1}{2(3\alpha\beta)^2} \left(2\Omega' + \dfrac{v'}{v}\Omega\right)^2, & v < 0 \end{cases} \tag{6.11}$$

which can be further simplified as:

$$\begin{cases} \gamma \left(2\Omega' + \chi'\Omega\right)^2 = \Omega^2 - 1, & v > 0 \\[2mm] \gamma \left(2\Omega' + \chi'\Omega\right)^2 = \Omega^2 + 1, & v < 0 \end{cases}$$

where we have introduced two new parameters:

$$\chi(x) = \ln |v(x)|, \quad \gamma = \frac{1}{2(3\alpha\beta)^2}. \tag{6.12}$$

It is at this point that we should fulfill the promise given at the end of a previous section and finally define the free parameter β. The idea here is to choose a number that yields a simplest possible coefficient γ. Naturally, we want $\gamma = 1$, so we define β as:

$$\beta = \frac{1}{3\sqrt{2}\alpha},$$

and end up with the following system of ordinary differential equations:

$$2\Omega' = \begin{cases} -\chi'\Omega \pm \sqrt{\Omega^2 - 1}, & v > 0 \\[2mm] -\chi'\Omega \pm \sqrt{\Omega^2 + 1}, & v < 0 \end{cases} \tag{6.13}$$

The system (6.13) can be further simplified by an appropriate choice of a new variable:

$$\Omega = \begin{cases} \dfrac{y}{\sqrt{y^2 - 1}}, & y' = -\dfrac{1}{2}(y^2 - 1)(-\chi'y \pm 1), & v > 0 \\[4mm] \dfrac{z}{\sqrt{1 - z^2}}, & z' = -\dfrac{1}{2}(z^2 - 1)(-\chi'z \pm 1), & v < 0 \end{cases} \tag{6.14}$$

The first thing we see in (6.14) is that both functions y and z satisfy the same equation that is called the *Abel equation of the first kind* [75]. However, the ranges of y and z are very different: $y = \{y \in \mathbb{R} : |y| > 1\}$ whereas $z = \{z \in \mathbb{R} : |z| < 1\}$. Thus, we end up with three important conclusions.

Conclusion 1. In the classical (non-phantom) case, the positively defined superpotential W satisfies the condition

$$W(x) = |v(x)| \cdot \Omega^2(x) = v(x) \cdot \theta(y(x)) = v(x)\, \frac{y^2(x)}{y^2(x) - 1}. \tag{6.15}$$

Conclusion 2. The function $y(x)$ is the solution of equation

$$\frac{dy}{dx} = -\frac{1}{2}(y^2 - 1)(\kappa - (\ln|v(x)|)'y), \tag{6.16}$$

where $\kappa = \pm 1$.

Conclusion 3. The region $|y| > 1$ corresponds to $v > 0$, and $|y| < 1$ corresponds to $v < 0$.

Before we conclude this section and begin our exploration of the equation (6.16) and its properties, it would be a good idea to look at the resulting graph of the newly constructed function $\theta(y) = y^2/(y^2 - 1)$ (see Fig. 6.1). First of all, we can see that this function has two vertical asymptotes at $y = \pm 1$. It also has a horizontal asymptote $\theta = 1$ that is only reached when $W = v$, and is therefore indicative of a vanishing kinetic term $1/2\dot{\varphi}^2$. But one thing that might look surprising is the inexplicable gap between the negative values of θ and its positive branches. What happens inside of a strip between $\theta = 0$ and $\theta = +1$?

In order to answer this question let us look back at the definition of W in (6.2). If we assume that θ, a proportionality factor between W and V,

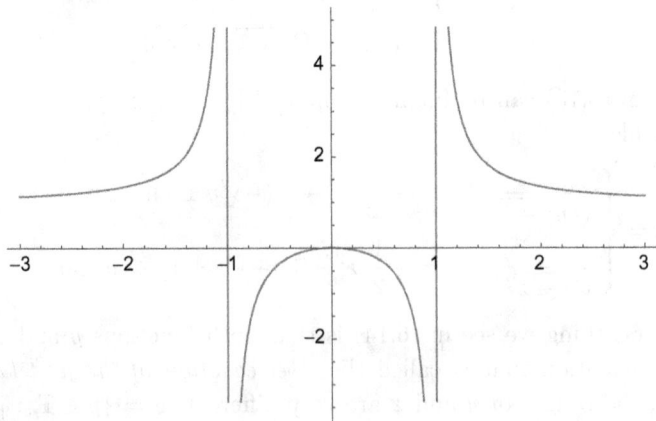

Fig. 6.1   A graph of function $\theta(y)$. It can be seen that the range of θ is $(-\infty, 0] \cup (1, +\infty)$.

hits a value inside of the strip $\theta \in (0,1)$, while still requiring that $W \geq 0$ (and thus $v > 0$ by (6.15)), we are essentially demanding the kinetic term to be negative, i.e.

$$\frac{1}{2}\dot{\varphi}^2 = W - V < 0, \tag{6.17}$$

so $0 < \theta < 1$ must describe the *phantom zone*! It is no wonder than that no such θ arose in our calculations.

What shall we do then if we are to account for the phantom zone in our formulas? The first thing we should do is acknowledge that for (6.17) to hold, the φ must be purely imaginary (the real part of φ must be equal to zero, or by (6.17), either the density $\rho = W$ or the potential V – or both! – will be complex-valued functions, which is clearly nonphysical). Therefore, one can introduce a new real-valued variable ψ such that:

$$\varphi = i\psi, \quad \psi \in \mathbb{R}. \tag{6.18}$$

Of course, the potential $V(\varphi) = V(i\psi) = \tilde{V}(\psi)$ must be real-valued for all ψ in the phantom domain, otherwise both the energy density ρ and the energy pressure p will be unphysically complex-valued (this automatically excludes all odd power functions like φ, φ^3, etc). Furthermore, the potential $\tilde{V}$ must be non-negative, otherwise we will end up with a negative energy density ($W < 0$ by (6.17)). This specifically implies that a standard massive field model $V_m(\varphi) = m^2\varphi^2/2$ lacks the prerequisites for existence of a phantom zone. On the other hand, both massless and massive scalar field models with a quartic interaction and corresponding potentials $V_\lambda(\varphi) = \lambda\varphi^4/4!$ and $V = V_m + V_\lambda$ can all lead to a phantom regime: the massless one for all $\psi \in \mathbb{R}$ and the massive one for all $|\psi| \geq 2m\sqrt{3/\lambda}$.

Remark 6.1. It is important to keep in mind that (the case of $V = \varphi^{4k}$, $k \in \mathbb{Z}$ notwithstanding) in general, the potentials V and $\tilde{V}$ have very different forms, properties and (real-valued) domains. For example, while the potential $V(\varphi) = \cosh(\varphi)$ is non-periodic, unbounded and positively defined for all real φ, its counterpart $\tilde{V}(\psi) = \cos(\psi)$ is periodic, bounded, and is only positively defined for $\psi \in (-\pi/2 + 2\pi k, \pi/2 + 2\pi k)$, $k \in \mathbb{Z}$.

Suppose we have the potential V that satisfies both of these conditions, that is:

$$\tilde{V}(\psi) = V(i\psi) : \mathbb{R} \to \mathbb{R}^+, \tag{6.19}$$

then instead of (6.4) we will have the following system:

$$\tilde{H}(\psi) = \pm \alpha \sqrt{\tilde{W}(\psi)},$$

$$\dot{\psi} = \pm \frac{1}{3\alpha} \frac{\tilde{W}'}{\sqrt{\tilde{W}}},$$

$$\tilde{V}(\psi) = \tilde{W}(\psi) + \frac{1}{2} \left(\frac{1}{3\alpha} \frac{\tilde{W}'(\psi)}{\sqrt{\tilde{W}(\psi)}} \right)^2. \tag{6.20}$$

Switching the variable ψ to x by setting $x = 3\sqrt{2}\alpha\psi$ and introducing a real-valued function Ω such that $\tilde{W}(x) = \tilde{v}(x)\,\Omega^2(x)$, we end up with the following equation:

$$1 = \Omega^2 + \left(2\Omega' + \frac{\tilde{v}'}{\tilde{v}} \Omega \right)^2. \tag{6.21}$$

This equation is remarkably similar to the first equation from (6.11); in fact, it *is* the same equation, just with an imaginary-valued parameter β.[2]

As before, we introduce a new function $\tilde{\chi} = \ln \tilde{v}$ and take the square root of Eq. (6.21); the result will be the differential equation

$$2\Omega' = -(\ln \tilde{v})'\, \Omega \pm \sqrt{1 - \Omega^2},$$

and so:

$$\Omega = \frac{y}{\sqrt{y^2 + 1}}, \quad y' = \frac{1}{2}(y^2 + 1)(\kappa - \tilde{\chi}'y). \tag{6.22}$$

Thus, the cosmological dynamics of a phantom zone is governed by the superpotential

$$\tilde{W} = \tilde{v} \cdot \Omega^2 = \tilde{v} \cdot \theta_p = \tilde{v}\, \frac{y^2}{y^2 + 1},$$

where y is the solution of equation

$$\frac{dy}{dx} = \frac{1}{2}(y^2 + 1)(\kappa - (\ln \tilde{v}(x))'y),$$

---

[2]It is curious to note how the different modes of the cosmological evolution seems to be controlled by a mere constant. However, as we have seen, the properties of the potential V should also be taken into account, as only those potentials that satisfy (6.19) will produce a physically meaningful solution for an imaginary β.

the variable x is defined as

$$x = -i3\sqrt{2}\alpha\varphi,$$

and $\tilde{v}(x) = V(\varphi)$.

The graphic of the function $\theta_p(y)$ is given in Fig. 6.2. Just as we expected, it fits right inside of the region $[0, 1)$ and therefore describes the dynamics of the universe filled with a phantom field. Therefore, we have considered all possible cases of the universes filled with a scalar field. The extended graph that combines both Figs. 6.1 and 6.2 is given on Fig. 6.3.

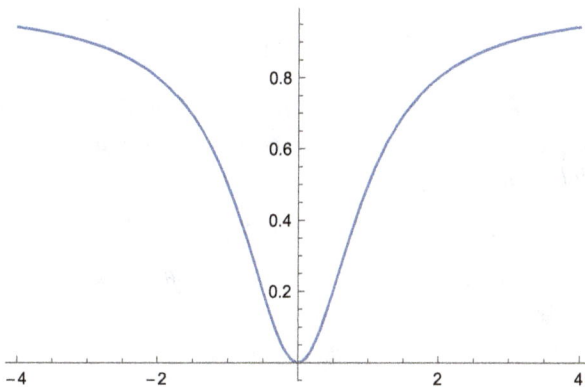

Fig. 6.2   A graph of function $\theta_p(y)$. The range of θ_p is $[0, 1)$.

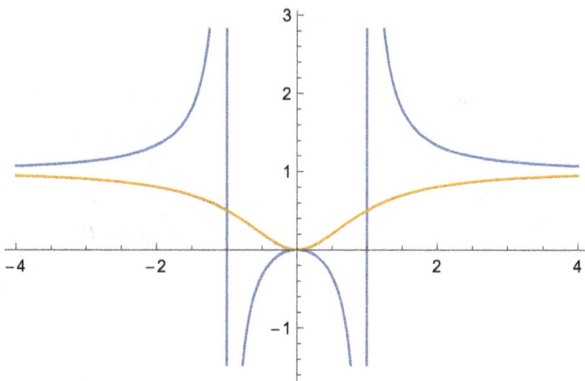

Fig. 6.3   A graph of both $\theta(y)$ and $\theta_p(y)$.

We are now prepared to conclude this section by compressing everything we have learned so far into the following Theorem (first appearance in [76]):

Theorem 6.1. *Let $x = 4\sqrt{3\pi}/M_P\,\varphi$, $\chi = \ln|V|$, $\kappa = \pm 1$. For a universe in a non-phantom zone with a given $V(\varphi)$ the corresponding superpotential $W = W(x, C)$ is defined as*

$$W(x, C) = V(x)\, \frac{y^2}{y^2 - 1}, \tag{6.23}$$

where $y = y(x, C) \neq \pm 1$ is a general solution of the Abel's equation of 1st kind:

$$y' = -\frac{1}{2}\left(y^2 - 1\right)\left(\kappa - \chi'y\right). \tag{6.24}$$

The positive values of V correspond to $|y| > 1$ and negative values of V correspond to $|y| < 1$. The case $V = 0$ occurs if and only if $y = \pm 1$ and the superpotential W has the form:

$$W = Ce^{\kappa x}.$$

Furthermore, if the potential $V > 0$ satisfies the condition $\tilde{V}(\xi) = V(i\xi) : \mathbb{R} \to \mathbb{R}^+$ for at least some ξ, for these values of ξ the universe can enter the phantom zone with superpotential W:

$$\tilde{W} = \tilde{V}\, \frac{y^2}{y^2 + 1}, \tag{6.25}$$

where $y : \mathbb{R} \to \mathbb{R}$ is the solution of the Abel equation

$$\frac{dy}{d\xi} = \frac{1}{2}(y^2 + 1)(\kappa - \tilde{\chi}'y). \tag{6.26}$$

Remark 6.2. (6.23) defines a family of solutions of (6.1), parameterized by the constant of integration C. Substituting $W(x, C)$ in (6.4) and solving it will produce $\varphi = \varphi(t; C, t_0)$, where t_0 is the second constant of integration, arising due to the invariance of the scalar field relative to the translations $t \to t + \text{const}$. Hence, by this process we will obtain *the general solution* of (6.1).

6.3 Abel equation at a glance

As we have seen in the previous section, the Friedman equations can be rather effortlessly transformed into a single ordinary differential equation of the first order, called the *Abel equation*, albeit with a twist: its exact form depends on whether or not the universe at a chosen time period undergoes the phantom fields-induced expansion. We will have ample time to discuss the peculiarities of the phantom zone in the corresponding chapter, so the remainder of this one will be dedicated to the classical dynamics and its corresponding Abel representation:

$$y' = -\frac{1}{2}\left(y^2 - 1\right)\left(\kappa - \chi' y\right). \tag{6.27}$$

The equation was originally introduced in a seminal work of Niels Henrik Abel [75] in 1881, and has seen a lot of extensive studies and applications ever since [77], [78] (see also [79], [80]). It often arises as a final product of a reduction of order for many high order systems of ordinary differential equations [31], [81] in various areas of mathematical physics. Naturally, this is not the first time the Abel equation arises in the studies of the Friedman-Robertson-Walker model. In fact, a part of our main Theorem 6.1 has been proven at least twice since the early 90s: by A.G. Muslimov in 1990 [16] (although the author did not recognize the resulting equation as the *Abel equation*) and then in the doctoral thesis of R.J.M. Easther (1993) [82]. Unfortunately, in both cases the proof was only made for the case of a strictly positive potential $V > 0$ and there has been no mention of the phantom scenario. Furthermore, both authors have treated their findings mostly as a mathematical curiosity and did not look for their deeper implications.

Let us now take a brief look at some of the important properties of Eq. (6.27).

First of all, we notice that the equation has two obvious stationary solutions: $y = \pm 1$. They correspond to $V = 0$ and produce the general solution of (6.1) in the form (see (6.1) for more details)

$$\varphi = \varphi_0 \mp \frac{\sqrt{2}}{3\alpha} \ln\left(1 + 3H_0(t - t_0)\right),$$

with the scale factor

$$a(t) = a_0\left(1 + 3H_0(t - t_0)\right)^{1/3}.$$

Thus, they describe the dynamics of a universe with a "stiff" equation of state $p = \rho$ (i.e., $w = 1$).

We are, of course, interested in the solutions that are non-stationary (and non-trivial). In order to find them it might be useful to rewrite Eq. (6.27) in a more handy manner. In general, the Abel equation of first kind is written as

$$y' = \sum_{\eta=0}^{3} f_\eta \, y^\eta, \tag{6.28}$$

where, in our case:

$$f_0 = \frac{\kappa}{2}, \quad f_1 = -\frac{1}{2}\chi', \quad f_2 = -\frac{\kappa}{2} = -f_0, \quad f_3 = \frac{1}{2}\chi' = -f_1. \tag{6.29}$$

It is known (cf., for example [80], [31]) that if f_1 is continuous, f_2 and f_3 are continuously differentiable and $f_3 \neq 0$ then one can represent Eq. (6.27) in a normal form:

$$\eta' = \eta^3 + J(x), \tag{6.30}$$

where

$$y = \omega(x)\eta(\xi) + \frac{\kappa V(x)}{3V'(x)}, \quad \omega(x) = \frac{1}{\sqrt{V(x)}} \exp\left(-\frac{1}{6} \int^x \frac{V(z)}{V'(z)} dz\right),$$

$$\xi = \frac{1}{2} \int \frac{V'(x)\omega^2(x)}{V(x)} dx, \quad J = \frac{2\kappa \left(9V'' - V\right) V^2}{(3V'\omega)^3}. \tag{6.31}$$

Example 1. For the popular cosmological model with a quadratic potential[3]

$$V(\varphi) = \frac{m^2\varphi^2}{2}, \tag{6.32}$$

one can use the substitution (6.31) for the $\varphi > 0$ ($x > 0$). In this case

$$\omega(x) = \frac{6}{mx} e^{-x^2/24}, \quad J(x) = \frac{\kappa m^3}{23328} x^4 \left(x^2 - 18\right) e^{x^2/8},$$

$$\xi = \frac{3}{2m^2} \left[\text{Ei}\left(1, \frac{x^2}{12}\right) - \frac{12}{x^2} e^{-x^2/12}\right], \tag{6.33}$$

where Ei is the exponential integral.

Example 2. In the case of a general power function potential

$$V = \frac{\lambda\varphi^n}{n} = \frac{\lambda x^n}{18^{n/2} n}, \quad n \in \mathbb{R}$$

---

[3]In quantum field theory this model describes noninteracting massive scalar particles.

with a positive coupling $\lambda > 0$, the Abel equation of 1st kind can be transformed into a particular case of the *Abel equation of 2nd kind*. For this
potential $\chi' = n/x$. Assuming $|y_0| > 1$ and sign(y_0) $= -\kappa$, let us consider
y as a variable and $x = x(y)$ as a solution under the question. Introducing
new function $P = P(y) = \kappa x(y) - ny$ and substituting it into (6.27), one
gets

$$PP' = F_1(y)P + F_0(y), \tag{6.34}$$

with

$$F_1(y) = -n - \frac{2}{y^2 - 1}, \qquad F_0(y) = -\frac{2ny}{y^2 - 1}.$$

Lets define

$$P = u(y) + F(y), \qquad F(y) = \int_0^y F_1(z)dz = -ny + \log\left|\frac{y+1}{y-1}\right|.$$

Then Eq. (6.34) will be reduced to

$$(u + F)u' = F_0. \tag{6.35}$$

Finally, if we assume that $|y_0| \geq 1$, then $F_0 \neq 0$ and one can introduce a
new independent variable

$$\xi = \int_0^y F_0(z)dz = -n\log\left(y^2 - 1\right), \qquad u(y) = \eta(\xi).$$

so that Eq. (6.35) will take a normal form

$$(\eta + F)\,\eta' = 1.$$

Another form of the Abel equation of 1st kind (6.27) can be obtained if
we know at least one of its exact solutions. We, of course, have not just one
but two: the stationary solutions $y = \pm 1 \equiv K$. Calculating the function

$$E(x) = \exp\int\left(3f_3K^2 + 2f_2K + f_1\right)dx = V(x)e^{-\kappa Kx},$$

and using the substitution

$$y = K + \frac{E(x)}{z(x)},$$

one gets

$$z' + \frac{\Phi_1}{z} + \Phi_2 = 0, \tag{6.36}$$

where

$$\Phi_1 = \frac{1}{4} \left(V^2\right)' e^{-2\kappa K x}, \quad \Phi_2 = \frac{1}{2} e^{-\kappa K x} \left(3KV' - \kappa V\right).$$

For the case

$$V = V_0 e^{\kappa K x/3},$$

$\Phi_2 = 0$, and Eq. (6.36) is exactly solvable. This example will be thoroughly examined in the next chapter.

Chapter 7

The Abel equation and the inflationary dynamics

7.1 The problem of inflation

In the previous chapter, we have demonstrated that an intricate relationship exists that connects together the Einstein-Friedman equations for the dynamics of a flat non-phantom universe filled with a scalar field, with a much simpler (albeit, generally speaking, nonintegrable) first order equation

$$y' = -\frac{1}{2} \left(y^2 - 1 \right) (\kappa - \chi' y), \tag{7.1}$$

called the Abel equation of the first kind. In this chapter we will turn our attention away from the more theoretical framework that we were so far confined to, and back to a specific and extremely important physical model, that plays a tremendous role in modern cosmology: the model of cosmic inflation.

Although the origins of the inflationary model can be traced as far back as 1917, when W. de Sitter found the first solution of the Einstein equations for a universe with a non-zero cosmological constant [83], it did not become a focus of universal interest until the early 1980s, when A. Linde proposed the chaotic inflation scenario [5, 84]. It was a development of a new inflationary universe scenario [3, 4], whose predecessors in particular have been the model introduced by A. A. Starobinsky [1, 85, 86] and the inflationary scenario of A. Guth [2, 87, 88].

For the purposes of this chapter, we will study the process of inflation on the simplest model of a homogeneous and isotropic universe filled with the real-valued scalar field φ, which has an already familiar form of two equations

$$H^2 + \frac{k}{a^2} = \frac{8\pi}{3M_P^2} \left(\frac{1}{2}\dot{\varphi}^2 + V(\varphi) \right), \tag{7.2}$$

and

$$\ddot{\varphi} + 3H\dot{\varphi} + dV(\varphi)/d\varphi = 0, \tag{7.3}$$

where $k = 0, +1, -1$ correspond to an either spatially-flat, closed or open Friedman universe, $a = a(t)$ is a scale factor, and H is the Hubble parameter (as usual, we use the unit system where $\hbar = c = 1$ and $G = M_P^{-2}$).

It is important to recall that in order to get a good approximation of the inflationary solution, it is quite customary to resort to the so-called *slow-rolling approximation*. It imposes the condition that the changes in a scalar field are sufficiently slow for the following inequality to hold: $\frac{1}{2}\dot{\varphi}^2 \ll |V(\varphi)|$. If we also assume that the scale factor $a(t)$ changes fast enough for $H^2 \gg \frac{k}{a^2}$, while the change of $\dot{\varphi}$ is further restricted by $\ddot{\varphi} \ll V'(\varphi)$, we will come to the conclusion that $H^2 = \frac{8\pi}{3M_P^2} V(\varphi)$, i.e. that the Hubble parameter has a weak time dependence, and that the scale factor increases by the law

$$a(t) \sim e^{Ht}.$$

Thus, the universe expands exponentially, and inflation takes place.

According to the assumption [89], an *exit from inflation* occurs when the slow-rolling approximation becomes invalid, i.e. when the kinetic part of the scalar field energy density $\frac{1}{2}\dot{\varphi}^2$ becomes comparable to $|V(\varphi)|$.

The slow-rolling condition serves to explain the occurrence of inflation. Unfortunately, in many models it appears to offer no easy natural exit from an inflationary regime. Many papers dedicated to the problem of inflation's end for the exactly solvable cosmological models employ the so-called "fine tuning", i.e. a model-specific accurate adjustment of the model parameters [26, 53, 54, 90–92].

Some examples of models that have a natural exit from inflation and do not require a fickle aid of a "fine tuning" have been provided in [92], although the potentials used therein do not necessarily correspond to any (currently known) particle theory. Besides, the definition of a universe evolution given there might be incomplete.

Thus, we end up with a problem that is an ideal subject on which we can test the power and versatility of our new method, described in the previous chapter! We are going to show how to use the reduction of the system (7.2), (7.3) for the purpose of analysis of the process of inflation in a spatially-flat Friedman universe filled with a real-valued scalar field with the potential

$$V(\varphi) = \frac{m^2\varphi^2}{2} + \frac{\lambda\varphi^4}{4}, \tag{7.4}$$

where $\lambda = 10^{-14}$, $m^2 = \lambda M_P^2$ according to Ref. [89]. The particular goals of our investigation can then be summed up as follows:

- to ascertain the possible necessary and sufficient conditions for validity of the slow-rolling approximation in the case of inflation with a natural exit;
- to find the initial values of a scalar field φ and its rate of change $\dot{\varphi}$, which would be necessary and sufficient to begin an inflationary phase during the evolution of a scalar field;
- to identify the effect that the ratio between the potential and kinetic terms of the energy density, as well as the initial values of φ, $\dot{\varphi}$, have on inflation phase, its time, and on the number of e-folds;
- to estimate the percentage of e-folds and the duration of the inflation over a period of a valid slow-rolling approximation.

7.2 A simple $m^2\varphi^2/2$ model: the inflation and slow-rolling condition

As we have mentioned before, the Abel representation (7.1) of the equations (7.2), (7.3) can be extremely useful even in those cases when one cannot find its exact solution. We will begin by considering one of the simpler yet popular cosmological models: the model of the universe filled with the scalar field with the quadratic potential

$$V(\varphi) = \frac{m^2\varphi^2}{2}. \tag{7.5}$$

It is easy to see that the Abel equation with such potential turns out to be non-integrable; nevertheless, as we shall see, it still functions as a very effective tool for studying the possible occurrence of inflation and the natural exit from it.

To see this, let us reintroduce the quantity $\theta(y)$, familiar to us from the previous chapter:

$$\theta(y) = \frac{y^2}{y^2 - 1} = 1 + \frac{1}{y^2 - 1}.$$

This quantity serves a role of a proportionality coefficient between the (unknown) superpotential W and (known) potential V:

$$W = \theta(y)V. \tag{7.6}$$

As we have seen in the previous chapter, the function $\theta(y)$ consists of three branches: the branch $\{\theta \geq 1, |y| > 1\}$, which corresponds to a non-phantom dynamics of positively defined potential V; the branch $\{\theta \leq 0, |y| < 1\}$, describing the non-phantom dynamics for negative V; and the intermediary branch $\{0 < \theta < 1, y \in \mathbb{R}\}$, responsible for the phantom zone dynamics. Since we are primarily interested in the classical inflationary dynamics, and owing to the fact that all potentials for the remainder of this chapter are to be positively defined, we will only work with the first branch:

$$\theta(y) \geq 1,$$
$$|y| > 1, \tag{7.7}$$

Let us now recall the definition of superpotential W:

$$W = \frac{1}{2}\dot{\varphi}^2 + V(\varphi).$$

Combining this with the condition (7.6), we end up with

$$\frac{\dot{\varphi}^2}{2} = V(\theta - 1). \tag{7.8}$$

On the other hand, the famous *slow-rolling condition* has the form:

$$\frac{\dot{\varphi}^2}{2} \ll |V|,$$

and in our framework it therefore assumes the form

$$\frac{V}{|V|}(\theta(y) - 1) \ll 1. \tag{7.9}$$

Since (7.5) requires $V \geq 0$, (7.9) is satisfied for any $|y| \gg \sqrt{2}$.

An inflation takes place whenever $\ddot{a}(t)/a(t) > 0$. The Friedman equation

$$\frac{\ddot{a}}{a} = -\frac{4\pi G}{3}(\rho + 3p),$$

tells us that this condition is identical to inequality $\rho + 3p < 0$, which can be simplified even further if we recall that the density ρ and pressure p satisfy the conditions

$$\rho = \frac{1}{2}\dot{\varphi}^2 + V(\varphi), \qquad p = \frac{1}{2}\dot{\varphi}^2 - V(\varphi).$$

Using these identities together with the condition (7.8) (and the fact that $V \geq 0$) will immediately reduce the inflationary condition to $\theta(y) < 3/2$ and, finally, to $|y| > \sqrt{3}$. The pressure p will be negative if $\theta(y) < 2$ or $|y| > \sqrt{2}$. Hence, we can determine the ranges for the values of y that would correspond to the different stages of the universe's evolution (Table 7.1).

Table 7.1 Prerequisites for the inflation for various values of y.

	Slow-rolling	Inflation	Negative pressure
I: $\sqrt{2} \ll y_* < y < \infty$	yes	yes	yes
II: $\sqrt{3} < y < y_*$	no	yes	yes
III: $\sqrt{2} < y < \sqrt{3}$	no	no	yes
IV: $y < \sqrt{2}$	no	no	no

Thus, inflation might actually continue even after the slow-rolling condition has been compromised, because when y crosses the range II, the inflation would still be maintained up to the point when y would finally reach $\sqrt{3}$.

7.3 The Abel equation of the 1st kind for the model $m^2\varphi^2/2 + \lambda\varphi^4/4$

Let us start by writing down the Abel equation of the 1st kind for a potential $m^2\varphi^2/2 + \lambda\varphi^4/4$ in form (7.1). (As before, we restrict ourselves to the case $k = 0$ for the spatially-flat universe.)

Following Theorem 6.1, after substitution

$$\varphi = \frac{x}{4\sqrt{3\pi}} M_P, \tag{7.10}$$

we get a potential (7.4) as a function of x:

$$V(x) = \frac{\lambda}{96\pi} x^2 \left(1 + \frac{1}{96\pi} x^2\right) M_P^4, \tag{7.11}$$

assuming $m^2 = \lambda M_P^2$, according to Ref. [89]. The Abel equation of the 1st kind, corresponding to Eqs. (7.2), (7.3), becomes

$$y' = -\frac{1}{2}\left(y^2 - 1\right)\left(1 - \left(\frac{2}{x} + \frac{2x}{96\pi + x^2}\right) y\right), \tag{7.12}$$

where the scalar field φ is directly proportional to x (see Theorem 6.1):

$$x = 4\sqrt{3\pi}/M_P\,\varphi$$

According to the Table 7.1 in the previous section, if we want the inflation to occur, the solution of the Abel equation (7.12) has to be greater than $\sqrt{3}$ for a certain interval of values of x. Since the scalar field is proportional to x and decreases during the inflation, the interval in question shall be firmly embedded inside of a larger interval $[0, x_0]$, where x_0 is the initial value of a scalar field.

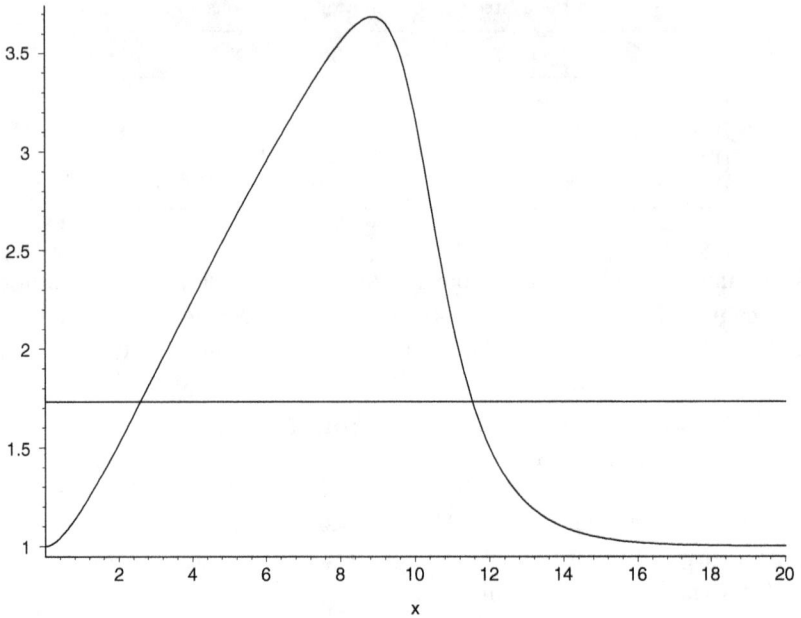

Fig. 7.1 The plot of the solution of the Abel equation (7.12) for the initial values $y(12) = 1{,}5$. The horizontal line signifies the inflationary condition $y = \sqrt{3}$.

As a first step, the minimal initial values $y_0 = y(x_0)$ were found, such that the corresponding solutions of the Abel equation (7.12) (Fig. 7.1) remained finite. The plot of maximal y_0 for corresponding x_0 still providing the finite solutions is on Fig. 7.2.

At this point the reader might ask: but why should we restrict ourselves to the regular solutions only? What is wrong with the singular solutions? The problem here lies not so much in the divergence of the solution $y(x)$, but in the consequences such a divergence will have for the dynamics of the universe. The cause for our concerns can be easily understood by recalling the very relationship between the superpotential W and the scalar field potential V (see (6.23)). When the solution to the Abel equation becomes singular, i.e. when $y \to \pm\infty$, according to (6.23) we must also have $W \to V$. This, of course, implies that $\dot{\phi}^2$ must vanish, which is a necessary prerequisite for the universe to *cross the phantom zone* (see Fig. 6.3). As we shall see in Part V, the discussion of such fascinating phenomena as the phantom zone crossing requires a whole new set of approaches, so it will be a good idea to postpone out

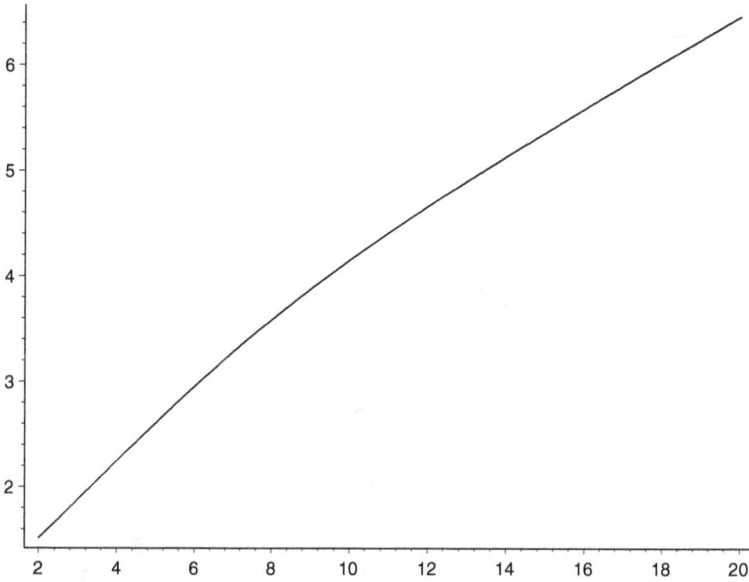

Fig. 7.2   The graph of highest possible y_0 that still yields regular solutions of the Abel equation (7.12) for given x_0.

investigation of the singular y's for later. Instead for the remainder of this chapter we will concentrate on the case of regular solutions of the Abel equation.

The solutions of Eq. (7.12) were derived by the computer algebra system Maple. The plots of y_0 for corresponding x_0 are displayed on Fig. 7.3. There one can see that an increase in x_0 leads to a decrease of a minimal y_0 necessary for inflation to occur. Up to $x_0 = 2.7$, there are no finite solutions of (7.12) that might exceed the value $y = \sqrt{3}$ in the range of x on the left of x_0.

7.4   The sufficient initial conditions for inflation to occur

Two parameters were chosen as the measures of the inflation: the logarithm of the ratio between the scale factor values at points $y(x_f) = \sqrt{3}$ $(x_f < x_i)$ and $y(x_i) = y_i$ $(y_i \geq \sqrt{3}, x_i \leq x_0)$, and the time interval separating these two moments. These parameters lead to the inflation provided the following estimates are true:

$$\ln \frac{a_f}{a_i} \gtrsim 100,$$

$$t_f - t_i \lesssim 1.9 \cdot 10^8 \text{ M}_{\text{P}}^{-1},$$

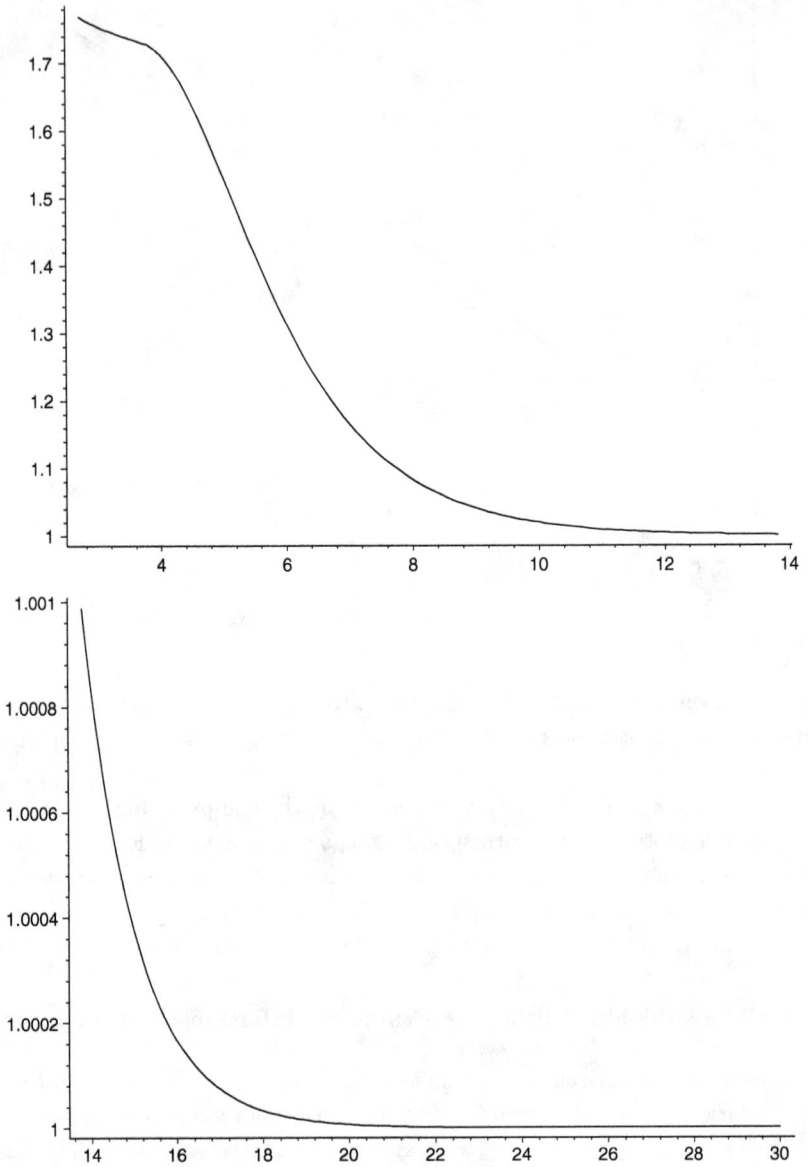

Fig. 7.3 The plot of minimal y_0, which provide finite solutions of the Abel equation (7.12), and for which these solutions are greater than $\sqrt{3}$ for some open interval on the x axis, for corresponding x_0.

where a_i, t_i are the values of the scale factor and time when $y(x_i) = y_i$, and a_f, t_f are, correspondingly, the value of the scale factor and the time when $y(x_f) = \sqrt{3}$.

To estimate $\ln \frac{a_f}{a_i}$ and $t_f - t_i$, we use the fact that $\frac{d(\ln a)}{dt} = H$. Recalling from the previous chapter that

$$H = \pm \frac{1}{M_P} \sqrt{\frac{8\pi}{3} W},$$

and using the formula (7.11), we get

$$d(\ln a) = H(x)\, dt = H(x)\frac{dt}{dx}\, dx = \pm\frac{1}{6}\sqrt{\frac{W(x)}{W(x) - V(x)}}\, dx = -\frac{1}{6}y(x)\, dx.$$

The choice of a minus sign is based on assumption that the universe is experiencing the inflation, thus making the Hubble parameter positive, and the rate of change of scalar field — negative.

Next,

$$\ln \frac{a_f}{a_i} = -\frac{1}{6}\int_{x_i}^{x_f} y(x)\, dx = \frac{1}{6}\int_{x_f}^{x_i} y(x)\, dx,$$

so, using (7.6) along with (7.11) yields

$$t_f - t_i = -\frac{M_P}{4\sqrt{\pi}}\int_{x_i}^{x_f} \frac{dx}{\sqrt{W(x) - V(x)}} = \frac{M_P}{4\sqrt{\pi}}\int_{x_f}^{x_i} \sqrt{\frac{y^2(x) - 1}{V(x)}}\, dx.$$

The consequent numeric integration produces the required estimate.

To find the minimal values (x_0, y_0), which are necessary for inflation to occur, we choose the following steps: $h_x = 0.1$ for x_0, $h_y = 0.001$ for y_0. It is determined that inflation can occur if the initial values are more than $x_0 = 65.2$, $y_0 = 16.573$, which correspond to the scalar field value $\varphi_0 = 5.3\, M_P$ and its rate of change value $\dot{\varphi}_0 = -1.2 \cdot 10^{-7}\, M_P^2$ for $\lambda = 10^{14}$. In this case, $\ln \frac{a_f}{a_i}$ and $t_f - t_i$ have the respective values 100.0 and $1.1 \cdot 10^8\, M_P^{-1}$. During the inflationary phase y_0 decreases sharply along with the increase in x_0, just as plotted on Fig. 7.4.

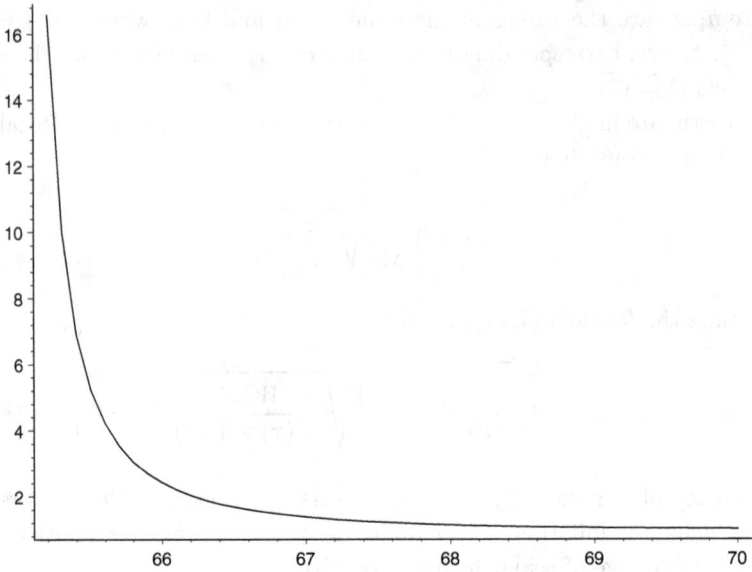

Fig. 7.4   The plot of a minimal $y_0(x_0)$, sufficient for an inflation to occur. ($\lambda = 10^{-14}$).

7.5   The dependence of the e-folds number and the time span of inflation upon the initial values (x_0, y_0)

The plots of the e-folds number and the inflationary time span are displayed on Figs. 7.5 and 7.6 respectively. One can notice that the e-folds number $P_{inf} = \ln \frac{a_f}{a_i}$ and time period $t_{inf} = t_f - t_i$ grow fast when $y_0 \gtrsim 1$ and increases. However, the bigger y_0 gets, the lesser its effect becomes. An increase in x_0 leads to a growth of both the e-folds number and inflationary time span. Also one can note that the dependencies of the P_{inf} and t_{inf} on y_0 decrease for large x_0. The values of the e-folds number and the time span of inflation for different x_0, y_0 are shown in Tables 7.2, 7.3. They contain the ratio between the inflation time and e-folds number for $y_0 = 10$, $y_0 = 20$, $y_0 = 30$ and $x_0 = 200$, $x_0 = 400$, $x_0 = 600$. From Table 7.2 it appears that doubling and tripling of x_0, with y_0 fixed, leads to the growth of the e-folds number which is more noticeable than that of an inflationary time span. The data in Table 7.3 shows that an increase in y_0 results in a slow growth of the e-folds number and a slower growth of the inflation's duration while x_0 is constant.

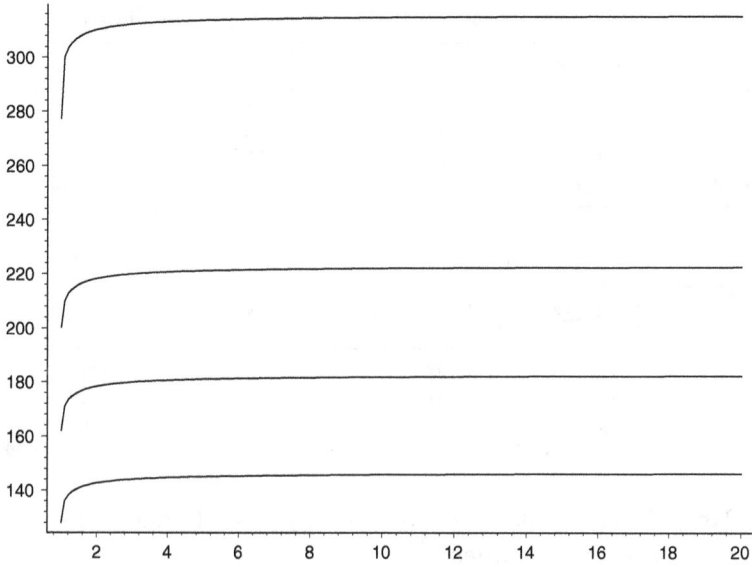

Fig. 7.5   The plot of the relation between the e-folds number and y_0 for different x_0 (top-down $x_0 = 120$, $x_0 = 100$, $x_0 = 90$, $x_0 = 80$).

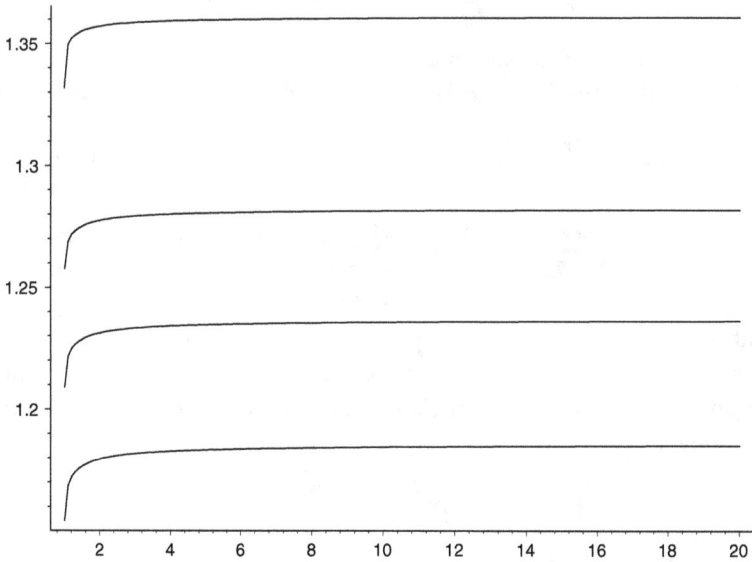

Fig. 7.6   The plot of the relation between the time of inflation $t_{inf} \cdot 10^{-8} \, M_P^{-1}$ and y_0 for different x_0 (top-down $x_0 = 120$, $x_0 = 100$, $x_0 = 90$, $x_0 = 80$).

Table 7.2   Ratios of the quantities $P_{\text{inf}}(x_0)$ and $t_{\text{inf}}(x_0)$ for different y_0.

y_0	$\dfrac{P_{inf(x_0=400)}}{P_{inf(x_0=200)}}$	$\dfrac{t_{inf(x_0=400)}}{t_{inf(x_0=200)}}$	$\dfrac{P_{inf(x_0=600)}}{P_{inf(x_0=200)}}$	$\dfrac{t_{inf(x_0=600)}}{t_{inf(x_0=200)}}$
10	3,941	1,19025	8,840	1,30152
20	3,940	1,19014	8,834	1,30138
30	3,939	1,19010	8,831	1,30132

Table 7.3   Ratios of the quantities $P_{\text{inf}}(y_0)$ and $t_{\text{inf}}(y_0)$ for different x_0.

x_0	$\dfrac{P_{inf(y_0=20)}}{P_{inf(y_0=10)}}$	$\dfrac{t_{inf(y_0=20)}}{t_{inf(y_0=10)}}$	$\dfrac{P_{inf(y_0=30)}}{P_{inf(y_0=10)}}$	$\dfrac{t_{inf(y_0=30)}}{t_{inf(y_0=10)}}$
200	1,0011	1,00015	1,0015	1,00021
400	1,0005	1,00006	1,0007	1,00008
600	1,0004	1,00004	1,0005	1,00005

7.6   How do the scalar field and its rate of change depend on the x and y values?

Expression (7.10) establishes a relationship between the scalar field and the x value

$$\varphi = \frac{x}{4\sqrt{3\pi}}\, M_P = 8{,}1 \cdot 10^{-2} \cdot x\, M_P.$$

The corresponding relationship between the rate of change of a scalar field and the x and y variables can be obtained from the Eq. (7.10) and (7.11):

$$\dot{\varphi} = -\frac{\sqrt{2\lambda}}{96\pi}\frac{x\sqrt{96\pi + x^2}}{\sqrt{y^2 - 1}}\, M_P^2 = -4{,}7 \cdot 10^{-10} \cdot \frac{x\sqrt{96\pi + x^2}}{\sqrt{y^2 - 1}}\, M_P^2, \qquad (7.13)$$

for $\lambda = 10^{-14}$, $m^2 = \lambda M_P^2$. We use the values $y > 1$ to resolve Eq. (7.12) in this case [76]. We choose the minus sign, since during the evolution the rate of change of the scalar field has to be negative. The example of its evolution during the inflationary phase is displayed on Fig. 7.7.

As a next step, let us turn our attention to the slow-rolling condition. More specifically, let us find out those y for whom the slow-rolling condition breaks down:

$$\frac{1}{2}\dot{\varphi}^2 \ll |V(\varphi)|$$

or

$$\frac{\dot{\varphi}^2}{2|V(\varphi)|} \ll 1. \qquad (7.14)$$

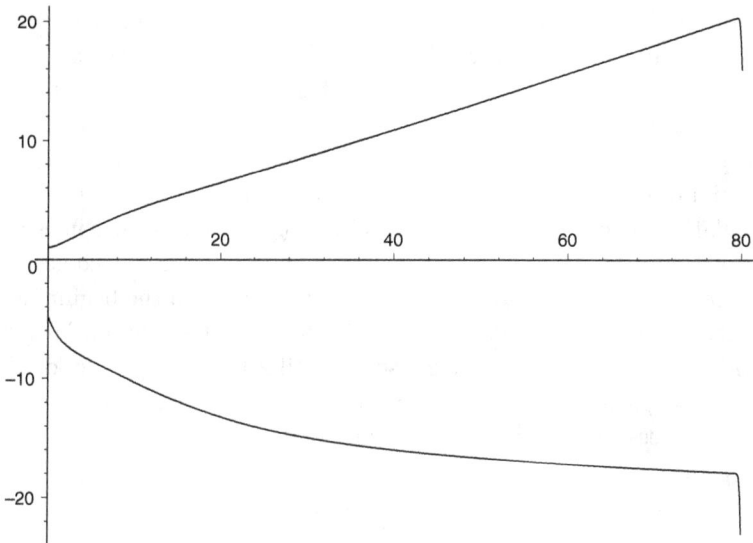

Fig. 7.7   The graphs of the solution of the Abel equation (7.12) (upper curve) for the initial conditions $y(80) = 16$ and the corresponding rate of change of the scalar field $\dot{\varphi} \cdot 5 \cdot 10^8 \ M_{\rm P}^2$ during its evolution (lower curve).

Under the assumption that $\frac{1}{2}\dot{\varphi}^2$ has to be at least ten times smaller than $|V(\varphi)|$, we determine that the slow-rolling condition is violated when $y \lesssim \sqrt{11}$.

Now, what can be said about an influence that the initial value of the scalar field φ_0 and initial ratio $2\frac{|V(\varphi_0)|}{\dot{\varphi}_0^2}$ have on the e-folds number and the inflation time? From the expression (7.6), we get

$$x_0 = \frac{4\sqrt{3\pi}}{M_{\rm P}}\varphi_0, \tag{7.15}$$

and from (7.14)

$$y_0 = \sqrt{2\frac{V(\varphi_0)}{\dot{\varphi}_0^2} + 1}. \tag{7.16}$$

In Section 7.5 we have shown that the increase in x_0 and y_0 leads to a growth of the e-folds number and inflation time, and the y_0 effect is insignificant when $y_0 \gtrsim 2$. Moreover, y_0, which is necessary for inflation to occur, decreases when x_0 grows. Thus, from the expressions (7.15), (7.16) it appears that the answers to the questions of whether or not the inflation commences, what would be its e-folds number and how long it will take for the inflation to run its course — all of them strongly depend on the initial value of the scalar field φ_0, and weakly — on the initial ratio between the potential and kinetic terms of the energy density of the scalar field. The e-folds number and inflation time increase with the initial value of the scalar field rising, according to the expression (7.15). An increase in the initial rate of change results in a decrease in y_0, and the e-folds number and inflation time diminish slowly. If $|\dot\varphi_0|$ is big enough, y_0 becomes of the order of 1, and P_{inf} and t_{inf} decrease considerably. Thus, the choice of the initial value of a scalar field has a strong impact on both the beginning and intensity of the inflationary phase, while the impact of the initial *rate of change* of φ remains comparatively small until $2|V(\varphi_0)|/\dot\varphi_0^2 \gtrsim 3$. So, $V(\varphi_0)$ grows while φ_0 increases, and after that bigger and bigger values of $|\dot\varphi_0|$ becomes necessary for a violation of this condition and an abrupt decrease of P_{inf} and t_{inf}.

The natural exit from inflation occurs when $y < \sqrt{3}$ [76].

7.7 How the initial values of a scalar field and the energy ratio affect the slow-rolling condition and the process of inflation

It is a safe claim that, should the scalar field become big enough, the slow-rolling condition would follow regardless of an initial ratio between the potential and kinetic terms of its energy density. In other words, the bigger the scalar field gets, the smaller an influence of the ratio ends up being. For example, for $x_0 = 100$ (which corresponds to $\varphi_0 = 8.05$ M_P) the slow-rolling condition fails when $2|V(\varphi_0)|/\dot\varphi_0^2 \lesssim 10^{-18}$, and for $x_0 = 200$ ($\varphi_0 = 16.1$ M_P) the failure occurs only when $2|V(\varphi_0)|/\dot\varphi_0^2 \lesssim 10^{-21}$ for $\lambda = 10^{-14}$. The plot on Fig. 7.8 shows minimal y_0 for corresponding x_0, which is necessary for solution of Abel equation (7.12) to be more than $\sqrt{11}$ on some range of values of x located on the left of x_0.

These values satisfy the slow-rolling condition. Figure 7.9 combines two plots of y_0 for corresponding x_0, for which solutions of Eq. (7.12) exceed $\sqrt{3}$ and $\sqrt{11}$.

From the figures one can notice that y_0 required for the slow-rolling condition to occur diminishes as x_0 grows, i.e. the bigger initial value of the scalar field imposes less restrictions on its initial rate of change.

However, although for $\lambda = 10^{-14}$ the slow rolling condition occurs when $\varphi_0 = 0.6$ M_P and $\dot\varphi_0 = -2.0 \cdot 10^{-8}$ M_P^2, the condition $\ln\frac{a_f}{a_i} \gtrsim 100$ requires $\varphi_0 = 5.3$ M_P and $\dot\varphi_0 = -1.2 \cdot 10^{-7}$ M_P^2, so even if the slow-rolling condition

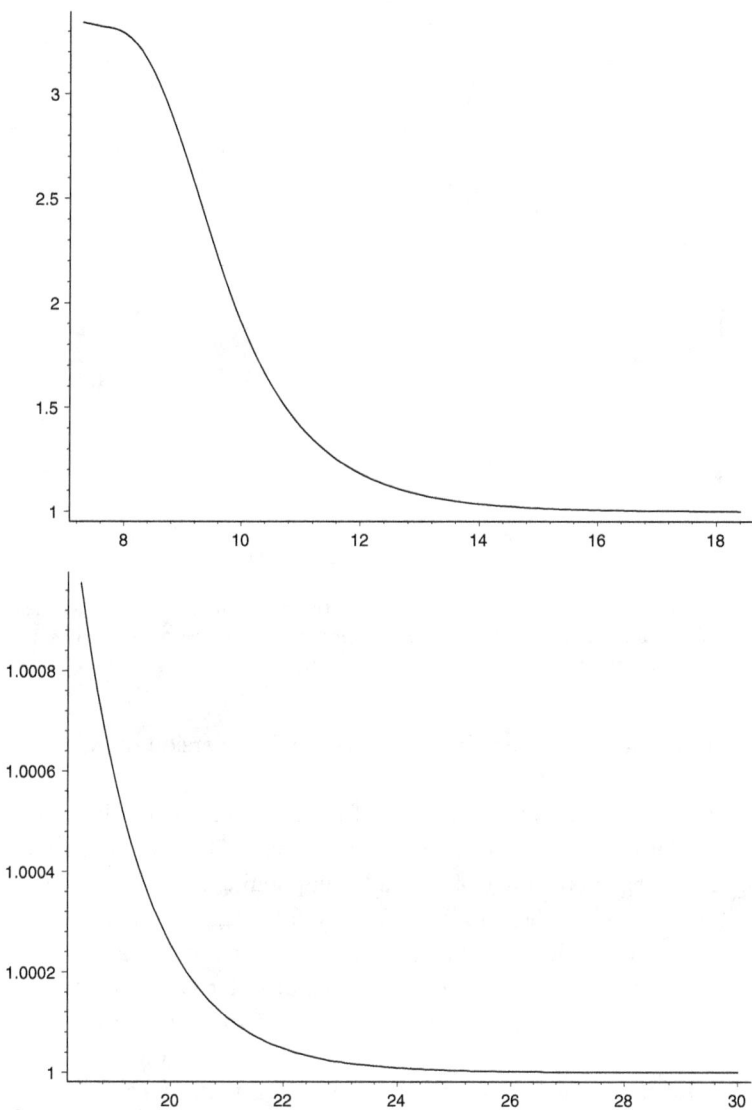

Fig. 7.8 The plot of minimal y_0, which provide finite solutions of the Abel equation (7.12), and for which these solutions are more than $\sqrt{11}$ on some range of the x axis, for corresponding x_0.

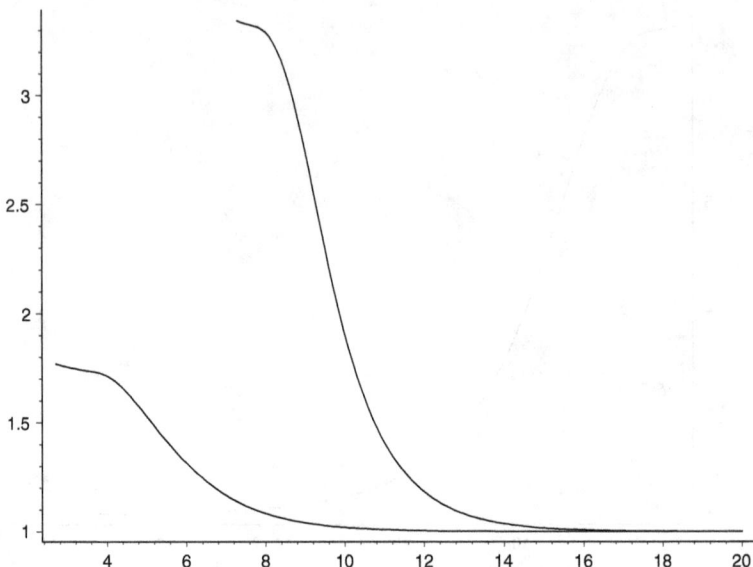

Fig. 7.9 The plots of the minimal $y_0(x_0)$ that (on some interval of x axis) yields the finite solutions of the Abel equation (7.12) bigger than $\sqrt{3}$ (lower curve) and $\sqrt{11}$ (upper curve), correspondingly.

rises but $\varphi_0 < 5.3$ M$_P$, we will only have a "mild" version of inflation with $\ln \frac{a_f}{a_i} \lesssim 100$.

In order to estimate the influence of the slow-rolling condition on inflation for different x_0, we have plotted the graphs of two relations: first, the one involving the percentage of all e-folds happening during the slow-rolling phase (as opposed to the total number of e-folds that is due to inflation in general), and the other one of the percentage of inflationary time spent during the slow-rolling phase. Both graphs are plotted as functions of x_0 for the value $y_0 = 0.001$ (Fig. 7.10 and Fig. 7.11 respectively).

On the figures, one can see that the slow-rolling condition is indeed satisfied during the bigger part of an inflationary phase. As a matter of fact, it lasts for approximately 82% of the inflationary phase and accounts for as much as 98% of the total e-folds number. If the initial value of the scalar field increases while the initial ratio between the potential and kinetic terms remains constant, those numbers become even bigger, although the actual growth during the inflation slows down. On the other hand, a comparable increase of the initial ratio between the potential and kinetic terms (with the constant initial value of the scalar field), would leave the percentages virtually unscathed, yielding only a minor increase in numbers (Fig. 7.12).

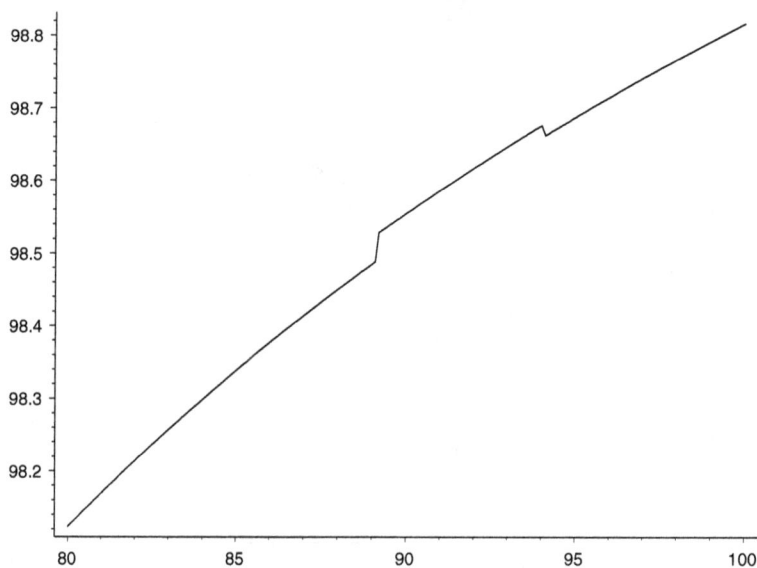

Fig. 7.10 The percentage of a number of e-folds arising during the slow-rolling phase as a function of x_0 for $y_0 = 1,001$.

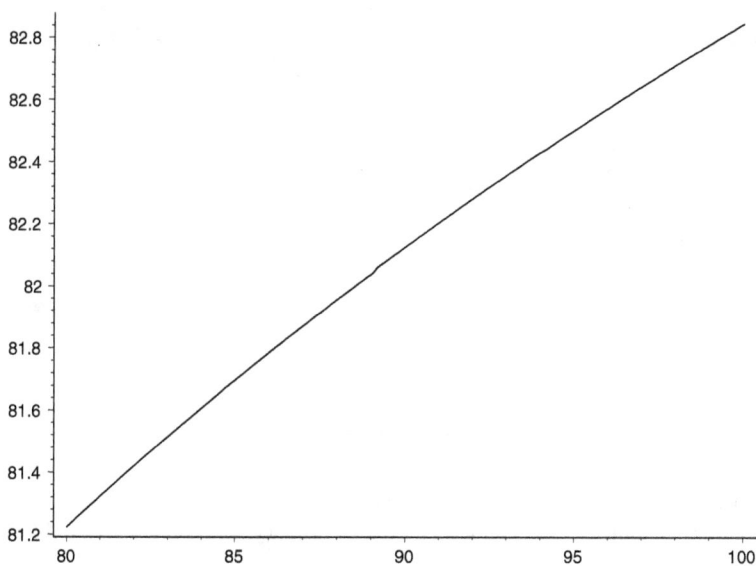

Fig. 7.11 The percentage of inflation's time spent during the slow-rolling phase as a function of x_0 for $y_0 = 1,001$.

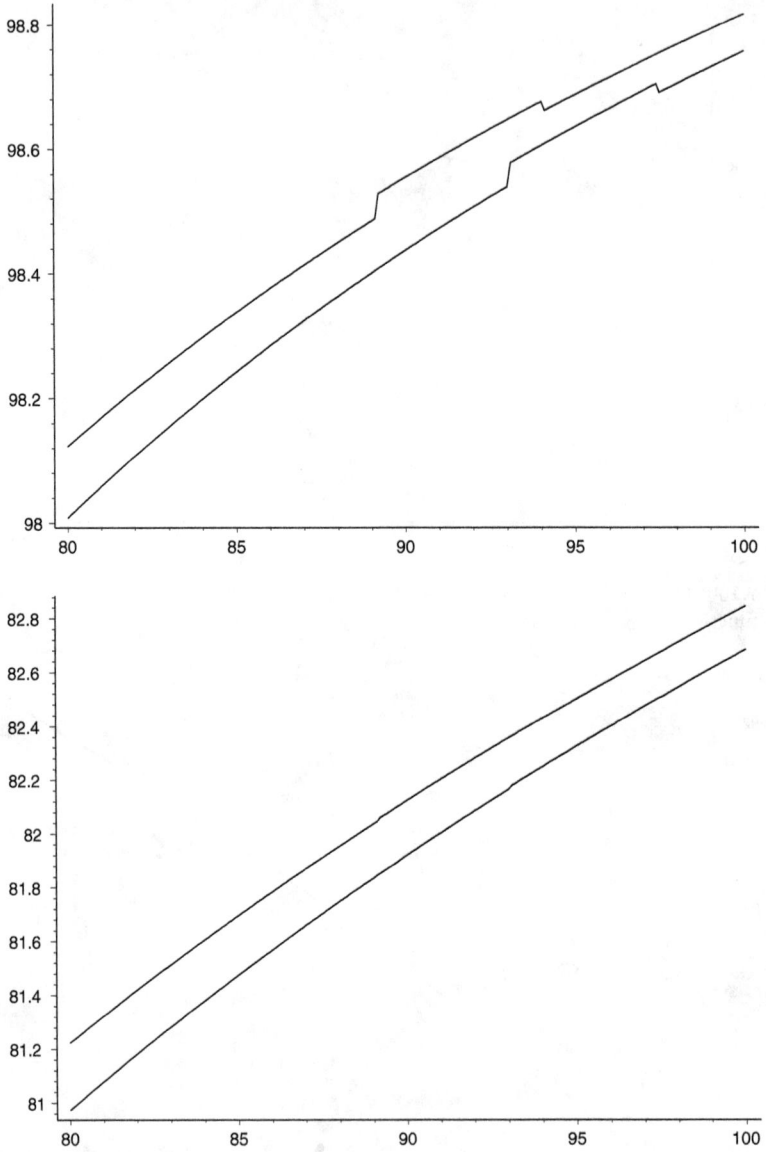

Fig. 7.12   The percentages of the e-folds number (left) and duration (right) of the slow-rolling phase as functions of x_0 for $y_0 = 1.25125$ (upper curve) and $y_0 = 1.001$ (lower curve)

7.8   The comparative study of $m^2\varphi^2/2+\lambda\varphi^4/4$, $m^2\varphi^2/2$ and $\lambda\varphi^4/4$ models

For a finishing touch, let us compare the inflationary dynamics of three models: $m^2\varphi^2/2 + \lambda\varphi^4/4$, $m^2\varphi^2/2$ and $\lambda\varphi^4/4$ (Table 7.4). The Abel equations describing the two latter models are:

$$y' = -\frac{1}{2}(y^2 - 1)\left(1 - \frac{2y}{x}\right), \tag{7.17}$$

$$y' = -\frac{1}{2}(y^2 - 1)\left(1 - \frac{4y}{x}\right). \tag{7.18}$$

The appropriate graphs of solutions of Eqs. (7.12), (7.17) and (7.18) can be seen on Fig. 7.13.

Table 7.4 shows that the inflationary dynamics in $m^2\varphi^2/2 + \lambda\varphi^4/4$ and $\lambda\varphi^4/4$ impose the lower bounds on initial values that are actually much higher than in $m^2\varphi^2/2$ case, thanks to the fact that for $\varphi \gg 1$, the scalar field of the latter changes much slower than it does for the first two. Moreover, it appears that the slow-rolling stage in $m^2\varphi^2/2$ accounts for a considerably bigger chunk of the inflationary phase than it does in the other two cases. The same is true for the e-folds number. Hence, although all three models yield the same e-folds number, the initial value of the scalar field

Table 7.4   The comparison of the results for $m^2\varphi^2/2 + \lambda\varphi^4/4$, $m^2\varphi^2/2$ and $\lambda\varphi^4/4$ models with $\lambda = 10^{-14}$, $m^2 = \lambda M_P^2$.

$V(\varphi)$	$\dfrac{m^2\varphi^2}{2} + \dfrac{\lambda\varphi^4}{4}$	$\dfrac{m^2\varphi^2}{2}$	$\dfrac{\lambda\varphi^4}{4}$
$V(x)$	$\dfrac{\lambda}{96\pi}x^2\left(1 + \dfrac{1}{96\pi}x^2\right)M_P^4$	$\dfrac{\lambda}{96\pi}x^2 M_P^4$	$\dfrac{\lambda}{9216\pi^2}x^4 M_P^4$
Minimal φ_0 and $\dot\varphi_0$, sufficient for inflation to occur:			
x_0	62.5	48.8	69.2
y_0	16.573	21.615	16.694
φ_0, M_P	5.3	4.0	5.6
$\dot\varphi_0$, M_P^2	$-1.2 \cdot 10^{-7}$	$-1.8 \cdot 10^{-8}$	$-1.3 \cdot 10^{-7}$
E-folds number	100.0	100.0	100.0
Inflation time t_{inf}, M_P^{-1}	$1.1 \cdot 10^8$	$2.2 \cdot 10^8$	$1.1 \cdot 10^8$
Minimal requirements on φ_0 and $\dot\varphi_0$ for the slow-rolling phase:			
x_0	7.3	6.2	12.8
y_0	3.343	3.361	3.339
φ_0, M_P	5.3	4.0	5.6
φ_0, M_P	0.6	0.5	1.0
$\dot\varphi_0$ M_P^2	$-2.0 \cdot 10^{-8}$	$-1.6 \cdot 10^{-8}$	$-2.4 \cdot 10^{-8}$

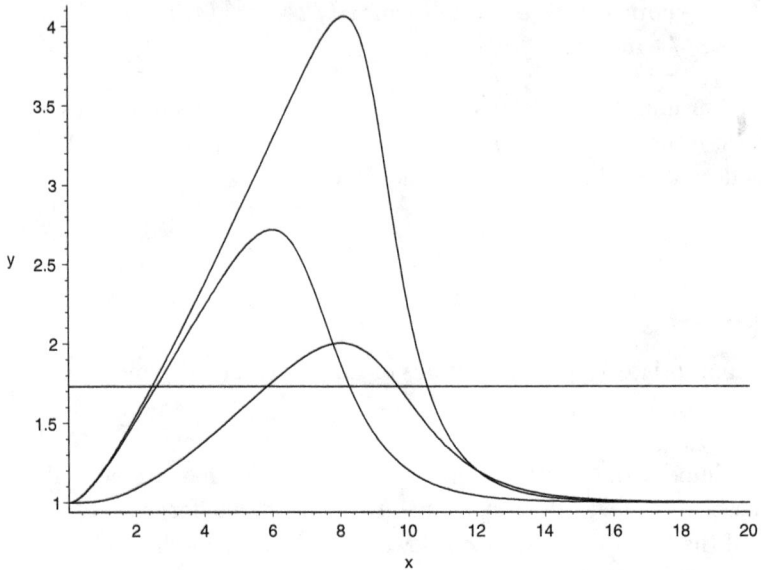

Fig. 7.13 The plots of the solutions of the Abel equations, corresponding to models $m^2\varphi^2/2 + \lambda\varphi^4/4$ (center curve), $m^2\varphi^2/2$ (upper curve), $\lambda\varphi^4/4$ (lower curve), for the initial value $y(12) = 1{,}5$

for the $m^2\varphi^2/2$ potential ends up being the smallest of the three, while the duration of inflation happens to be the longest. The models $m^2\varphi^2/2 + \lambda\varphi^4/4$ and $\lambda\varphi^4/4$ are quite similar to each other, but in the first case, the e-folds number associated with the slow-rolling condition is bigger, and the minimal φ_0 necessary for inflation to occur is smaller than in the second case.

As for the minimal initial rate of change of the scalar field, it is only natural for its absolute value to decrease along with the initial value of the scalar field, since the small value of $2|V(\varphi_0)|/\dot\varphi_0^2$ is more favorable for the inflation *per se*.

7.9   Discussion of the Results

We have devoted this chapter to the goal of demonstrating how the existing connection between the Einstein–Friedman equation (7.2), scalar field equation (7.3) and the Abel equation (7.1) can be used as a primary means of analysis of even most complex non-integrable cosmological models. In order to to do that, we have applied the Abel equation to the study of the inflationary dynamics in three particular models, each one describing a spatially-flat homogenous isotropic universe filled with scalar field φ, albeit with three different potentials: a massive field $V = m^2\varphi^2/2$, a massless quartic field $V = \lambda\varphi^4/4$ and the most general case of a massive field with a quartic interaction $V = m^2\varphi^2/2 + \lambda\varphi^4/4$. And, lo and behold! By sticking almost exclusively to the Abel equation, we were able to obtain a number of nontrivial results. In particular, we have found that:

- in most models the slow-rolling condition arises naturally during the dynamics of the scalar field; the necessary criterion for its absence is the small initial ratio between the potential and kinetic terms of the scalar field energy density; the growth of the initial value of the scalar field leads to a decrease of a necessary ratio; for example, if $\varphi_0 = 8.5\ M_P$ then the slow-rolling condition cannot occur when $2|V(\varphi_0)|/\dot{\varphi}_0^2 \lesssim 10^{-18}$;
- the e-folds number and the time span of inflation both grow with the increase in the initial value of the scalar field and/or initial ratio between the potential and kinetic terms, although the time span grows noticeably slower; however, the main influence on the process of the inflation has to be attributed to the initial value of the scalar field;
- if $|\dot{\varphi}_0|$ increases while φ_0 is constant then the e-folds number and the time of inflation decrease, however, if φ_0 is big enough, the change would be negligible; the effect becomes more noticeable if $|\dot{\varphi}_0|$ is sufficiently large and $2|V(\varphi_0)|/\dot{\varphi}_0^2 \lesssim 3$;
- the ratio $2|V(\varphi_0)|/\dot{\varphi}_0^2$ has little influence on the process of inflation while $2|V(\varphi_0)|/\dot{\varphi}_0^2 \gtrsim 3$; thus, if φ_0 is big enough, the restriction on φ_0 can be relaxed, and as φ_0 grows, so does the highest possible value of $|\dot{\varphi}_0|$;
- the condition $2|V(\varphi)|/\dot{\varphi}^2 > 2$ is necessary to initiate the inflation, while the natural exit from inflation requires $2|V(\varphi)|/\dot{\varphi}^2 < 2$, i.e. the inflation begins well before the slow-rolling phase and ends some time after;
- the bigger part of the e-folds number and the time span of the inflation falls on the period of the slow-rolling condition (about 98% for the e-folds and 82% for the time); these percentages grow with the initial value of the scalar field and the initial ratio between the potential and kinetic terms of its energy density, although the former imposes bigger influence than the latter.

Furthermore, these remarkable results were not only easy to gain, but they are also on a complete accord with the earlier estimates [89]. This confirms the reliability of the analysis method suggested in Ref. [76] and shows its potential for usage in other, more complicated models.

Part V
The phantom fields

Chapter 8

Phantom energy and symmetric cosmological solutions

8.1 Phantom quintessence

One of undeniable features of the contemporary cosmology is the consistency with which it produces new ideas that immediately become subjects of a vehement scientific contention. Among the most recent culprits are such concepts as big bang, black holes, inflation, cosmic strings, wormholes and, of course, the most recent duo of the accelerating expansion of the universe and the phantom energy. Curiously and most untraditionally, the blame for a concoction of the last two squarely lies not on the theoreticians, but on the astronomers and the observational data they came up with [93]. As a result, the controversy that surrounds the phantom energy is not so much a product of disbelief in its existence, but a reaction to the physical models that are proposed to explain it, and also to the fascinatingly counterintuitive properties of all of those models (negative pressure?.. Negative parameter of state?.. Scale factor having a *discontinuity*?..). Still, according to a famous anecdotal exchange between the Pauli and Heisenberg, sometimes to get a most accurate theory or interpretation one should ask not whether it is too crazy or unusual, but instead query whether it is crazy *enough*. Or, to quote the one and only consulting detective, Sherlock Holmes himself, "Once you eliminate the impossible, whatever remains, no matter how improbable, must be the truth" [94]. This is why in this chapter we will turn our attention to the three very unusual cosmological scenario, that, despite their strangeness, actually appears to be compatible with the contemporary observations. These three concepts are: the big rip [72], the big trip [95] and the big hole. As we shall see, these three "biggies", together with the *big bang* and big crunch, constitutes five of the most dramatic events that can theoretically take place in the universe.

Before we begin, however, it would be a good idea to briefly recap the theoretical reasoning that led to the introduction of the concept of *"phantom energy"* as well as the conceptual problems it produced. As we have already mentioned a number of times, the observational data gathered since 1999 unequivocally confirms that the universe at present is undergoing a period of accelerating expansion [96]. This poses the question of whether or not this acceleration can be wholly attributed to the influence of the cosmological constant; in mathematical terms this means checking whether the parameter w of the equation of state $p = w\rho^1$ is exactly equal to -1 — or not. So far the observations were understandably uncertain on this regard, showing only that w remains very close to, but perhaps a bit smaller than -1 [93]. Nevertheless, no matter how close to -1 it is, if $w < -1$ then our universe would be filled with a sort of a matter that is usually called a "phantom energy", ultimately serving to super-accelerate the expansion of the universe. Now, admittedly, the phantom energy is a curious, if not downright weird stuff [97]. While it has its share of beneficial properties — producing the observed accelerated expansion and, possibly, the predicted primordial inflation [98], to name but a few, — it is also known to possess a slew of stupefyingly unusual characteristics. In particular, if we describe it via a scalar field ϕ with the FLRW customary definitions, $\rho = \dot{\phi}^2/2 + V(\phi)$, $p = \dot{\phi}^2/2 + V(\phi)$, with ρ and p being the energy density and pressure, respectively, and $V(\phi)$ the field potential, then: (i) the kinetic term $\dot{\phi}^2/2 < 0$ and therefore the phantom cosmologies shall suffer some very violent instabilities, (ii) the energy density ends up being an increasing function of time, with the later stages of the cosmological evolution being dominated by a quantum-gravity regime, (iii) the *dominant energy condition* gets violated: $\rho + p < 0$, and (iv) there will be a singularity in the finite future dubbed the "big rip", when the universe effectively would cease to exist, this catastrophic event being preceded by the breakage of the causality on a cosmic scales. These properties define models of the phantom energy in a so-called *"quintessence scenario"*.

In order to understand the quintessence scenario and its generalizations, we must first understand what are the big rip, big trip and the big hole in the usual quintessence when the equation of state is $p = w\rho$. Using the general expression

$$\frac{\dot{\rho}}{\rho} = -3H(1 + w) = \frac{2\dot{H}}{H}, \tag{8.1}$$

---

[1]We are reminding our reader that we are using the system of units where $8\pi G/3 = c = 1$.

where $H = \dot{a}/a$ and w is assumed to be constant, we can derive the expressions for the scale factor $a(t)$ and the energy density in the quintessence model to be

$$a(t) = \left[a_0^{3(1+w)/2} + \frac{3}{2}C(1+w)(t-t_0) \right]^{2/[3(1+w)]}, \qquad (8.2)$$

in which C is a constant with a sign determined by the requirement $\dot{a} > 0$ (the universe must expand), and

$$\rho(t) = H^2 = \frac{C^2}{\left[a_0^{3(1+w)/2} + \frac{3}{2}C(1+w)(t-t_0) \right]^2}. \qquad (8.3)$$

Now, suppose the dynamics of the universe is governed by the phantom regime with $w < -1$. Then it can be seen that ρ monotonously increases with t up to the time

$$t = t_* = t_0 + \frac{2a_0^{-3(|w|-1)/2}}{3C(|w|-1)}, \qquad (8.4)$$

at which it diverges. It can be seen that the scale factor $a(t)$ also blows up at $t = t_*$. This curvature singularity is what we call *the big rip* [72]. If there exist no space-time branches of the FLRW solutions that can sew together the super-accelerating expanding region before the big rip to a contracting region afterward, then the big rip would mark the ultimate end of both the universe and everything in it. In the quintessence model the phantom energy will therefore be characterized by $\rho > 0$, $\rho + p < 0$, $\dot{\phi}^2 < 0$, a big rip singularity at $t = t_*$ and a positively defined potential

$$V(\phi) = \frac{1}{2}(|w| + 1)C^2 e^{-3i\sqrt{|w|-1}(\phi-\phi_0)}, \qquad (8.5)$$

with

$$\phi(t) = \phi_0 - \frac{2i}{3\sqrt{|w|-1}} \log \left[\left(\frac{a_0}{a} \right)^{3(|w|-1)/2} \right], \qquad (8.6)$$

in which ϕ_0 is another constant.

This is admittedly a very discomforting scenario for the proposed end of the universe; it is even more depressing than the frozen starless wasteland of a heat death for a perpetually expanding universe, or the fiery furnace of the Big Crunch singularity in the closed universe models. At least the latter two provides some few billions of years of respite! No such luck in the phantom scenario; here the big rip singularity is something that might happen literally in a matter of seconds (depending on the choice of constants a_0, t_0 and C). However, one should not despair too eagerly, for the very thing that makes the big rip so dangerous for the universe might ultimately protect its inhabitants! The saving throw in this case comes from a simple observation: that the phantom energy appears to be *exactly* a sort of the matter required for the formation of the stable Lorentzian *Morris-Thorne wormholes* [99]. Such wormholes therefore should be expected to copiously crop up in a universe dominated by a phantom energy. Once they are formed with a size sufficiently small to make them stable against the vacuum polarization, the *wormholes* would begin accreting the phantom energy to induce a swelling process in the wormhole spacetime, inflating the wormholes' throat so quickly that, relative to an asymptotic observer, they would engulf the entire universe by the time [95]:

$$\tilde{t} = t_0 + \frac{1}{\beta + \frac{3}{2}Ab_0(|w| - 1)}, \qquad (8.7)$$

where b_0 is the initial radius of a spherical wormhole throat, β is a constant and A is a numerical coefficient of order one. As a result, the universe would find itself passing through the "tunnel" separating one mouth of the wormhole from another right before the big rip, and, depending on the relative kinematic characteristics of the resulting insertions of the two wormhole mouths, the universe in question might in process travel in time towards the moment that lies well *beyond* the big rip singularity, effectively *skipping* it altogether. This scenario has been dubbed *the big trip* [95] and it appears to be the only sort of a cosmic-scale phenomena that really offers a possibility of escaping from the doomsday of a big rip.

Finally, let us discuss what we mean by a "big hole" and what it has to do with the phantom energy. It is commonly accepted today that our universe must contain a multitude of *black holes* with sizes ranging from a few solar masses to several billions of solar masses. The latter *black holes* are called supermassive and are thought to be located primarily in the galactic centres. Any such *black hole* accretes the dark energy with $w > -1$ in a process that parallels the phantom energy accretion by the wormholes and that might in principle lead to a swelling of their event horizon to such gigantic proportions that the black hole would eventually swallow the universe as a whole at time [100]

$$t_{bh} = t_0 + \frac{1}{\left[\frac{3}{2}B\rho_0 M_0 - (6\pi\rho_0)^{1/2}\right](1 + w)}, \qquad (8.8)$$

where M_0 is the initial mass of the black hole, ρ_0 is the initial energy density and B is a numerical constant of order one. Curiously, in the framework of the standard quintessence models this particular growth turns out to be irrelevant. In fact, putting reasonable astronomical data in (8.8) leads to conclusion that the accretion of a dark energy will significantly modify the black hole size prior to big rip only if the initial black hole mass M_0 would be of the order of the total mass of the universe, a situation which can never be expected to occur during the entire evolution of the universe [100]. However, this conclusion will no longer be applicable once we leave the standard quintessence scenarios and move on to the more general models. In fact, we will show that in these generalized models the *black holes* can and will experience a big hole phenomenon.

With all that being said, we are now fully prepared to move on to a detailed discussion of two FLRW solutions that will serve as a foundation for our discussion: the solutions with two big rips. As we will see, in order to do this the solutions must feature a very interesting type of matter which is effectively dual to the standard phantom energy.

8.2 Cosmological solutions with two big rip singularities

Let us start by considering a solution of FLRW equations with the following scale factor

$$a(t) = \alpha \left(\beta + x \tan x \right), \tag{8.9}$$

where $x = \kappa t$, κ, α and β are all positive constants. We will study this solution on the interval $-t_R < t < t_R$, where $t_R = \pi/2\kappa$. We shall hereafter refer to the scale factor (8.9) as *Solution I*. On Fig. 8.1 we provide a plot of such a solution which can be obtained by using the *Darboux transformation* [92].

It is easy to see that $a(\pm t_R) = +\infty$, i.e. that there are two big rip singularities at $t = \pm t_R$. Moreover, on the subinterval $-t_R < t < 0$ the universe undergoes a contraction that lasts until $t = 0$, after which the expansion takes over only to end in yet another big rip at $t = +t_R$. It is important to note here that $a(t)$ does not vanish at any time along the considered interval; the minimum value of the scale factor appears to be $a(0) = \alpha\beta$.

The described behavior of the universe, while highly unusual, nevertheless appears to be compatible with the observable universe, thanks to the courtesy of the free parameters in the definition of the scale factor (8.9).

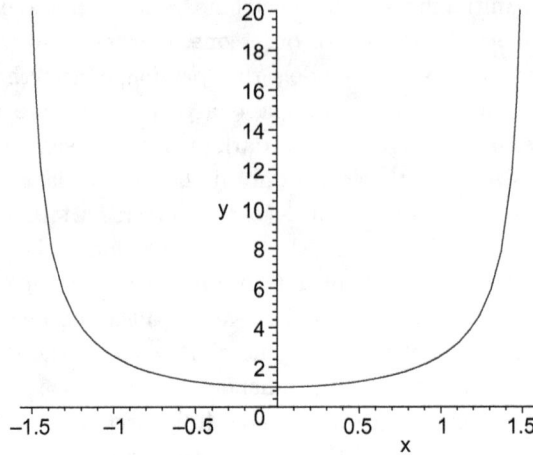

Fig. 8.1   The evolution of Solution I from one big rip situated in the past to another big rip located in the future.

In order to prove this, it will suffice to demonstrate that it is possible to choose the parameters in such a fashion that

(1) The solution will reproduce the observed value of the Hubble constant,
     $H(t_0) = H_0 = h \times 0.324 \times 10^{-19}$ s^{-1}, where $45 < h < 75$;
(2) $\ddot{a}_0/a_0 = 7H_0^2/10$;
(3) $a_0 \sim 10^{28}$ cm.

To show these simply substitute the solution (8.9) into the FLRW equations for a flat universe to get the energy density

$$\rho = \frac{\kappa^2(T + x + xT^2)^2}{(\beta + xT)^2} \tag{8.10}$$

and the pressure

$$p = -\frac{5\kappa^2 x^2 T^4 + 2x(2\beta + 3)T(T^2 + 1) + (4\beta + 6x^2 + 1)T^2 + 4\beta + x^2}{3(\beta + xT)^2}, \tag{8.11}$$

where $T = \tan x$. We can then obtain the expression for the Hubble constant

$$H = \frac{\kappa(2x + \sin(2x))}{2\cos x(\beta \cos x + x \sin x)} \tag{8.12}$$

and

$$\rho + 3p = -\frac{4\kappa^2(1 + xT)}{(\beta + xT)\cos^2 x}. \tag{8.13}$$

Since the combination given by Eq. (8.13) is negatively defined we deduce that the *strong energy condition* is permanently violated and therefore the universe must experience constant acceleration.

Let us now consider condition (2). We can write for the constant β

$$\beta = \frac{(80\cos^2 x_0 - 52)x_0^2 - 12x_0\sin(2x_0) + 7\sin^2(2x_0)}{80\cos x_0(x_0\sin x_0 + \cos x_0)}. \tag{8.14}$$

It can be seen that for small values of x_0, β behaves like an inverted *Higgs potential* (Fig. 8.2). The values of interest here would be those that lead to a positive β. Such values are bounded by $x_0 < x_m \sim 0.6$. We can then evaluate the constant κ to be

$$\kappa = \frac{7H_0\cos x_0(2x_0 + \sin(2x_0))}{40(x_0\sin x_0 + \cos x_0)}, \tag{8.15}$$

which in turn implies that

$$t_0 = \frac{x_0}{\kappa} = \frac{40x_0(x_0\sin x_0 + \cos x_0)}{7H_0\cos x_0(2x_0 + \sin(2x_0))} \tag{8.16}$$

and $t_R = \pi/(2\kappa)$. We can now calculate the length of time

$$\Delta t_R = t_R - t_0 = \frac{(\pi - 2x_0)t_0}{2x_0}. \tag{8.17}$$

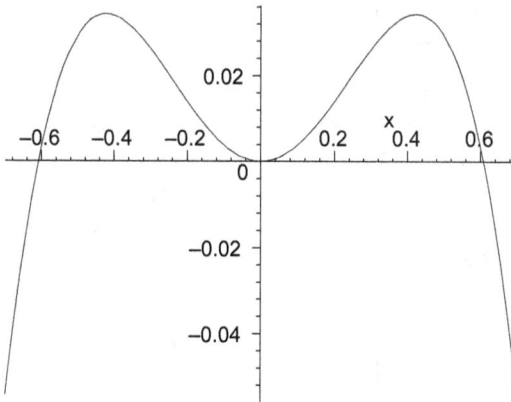

Fig. 8.2   The evolution of parameter β with time. At small x_0 it adopts the shape of an inverted Higgs potential.

The results of the numerical computations are summarized in the table. In the first row we give the different values of x_0 from 0.1 up to 0.6; in the second row — the least values of the scale factor, reached at $t = 0$ during the transition between two regimes of expansion. Next two rows present the values of t_0 calculated for two limiting values of the Hubble constant: $h = 45$ and $h = 75$, correspondingly. Those values are related to the time elapsed since the beginning of the expansion of the universe up to present time. The values of Δt_R for $h = 45$ and $h = 75$ are presented in the fifth and six rows. The last two rows contain the values computed for the age of the universe with respect to $h = 45$ and $h = 75$, respectively, i.e. the time passed from $t = -t_R$ to $t = t_0$. All times in the Table are expressed in standard Earth years.

x_0	0.1	0.2	0.3	0.4	0.5	0.6
a_{min}	$0.27a_0$	$0.26a_0$	$0.22a_0$	$0.17a_0$	$0.1a_0$	$0.01a_0$
$t_0/10^9$ (yr), $h = 45$	31.4	32.7	35	38.3	43	49.3
$t_0/10^9$ (yr), $h = 75$	18.9	19.7	20.1	23	25.8	29.6
$\Delta t_R/10^9$ (yr), $h = 45$	463	225	148	112	92	80
$\Delta t_R/10^9$ (yr), $h = 75$	278	135	89	67	55	48
$T/10^9$ (yr), $h = 45$	526	244	218	189	178	178
$T/10^9$ (yr), $h = 75$	316	170	130	113	107	107

At this point the reader might inquire: "All right then, this particular solution indeed seems to be compatible with the observations. But what about the *physics* of it?.." Well, it appears that there indeed exist a physical model that behaves in a fashion quite similar to our Solution I. This model in question is the *Randall-Sundrum brane world model* I [101] where the Friedmann equations can be written as

$$\frac{\dot{a}^2}{a^2} = \rho \left(1 + \frac{\rho}{2\lambda}\right), \qquad (8.18)$$

$$-2\frac{\ddot{a}}{a} = \rho + 3p + \frac{\rho}{\lambda}(2\rho + 3p). \qquad (8.19)$$

The corresponding solution — we will call it *Solution II* to distinguish it from the previously discussed Solution I — of the system (8.18), (8.19) has the form

$$a^{6\epsilon} = \frac{s}{1 - 18\lambda\epsilon^2 (t - t_1)^2}, \qquad (8.20)$$

where

$$\frac{p}{\rho} = w = -1 - 2\epsilon,$$

$\epsilon > 0$, $t_1 = $ const and $s = $ const.

Obviously, in order to have the big rip singularities one simply has to choose $\lambda > 0$. On the other hand, $a(t)$ must be positive so that $s > 0$. We have two big rips at

$$t_\pm = t_1 \pm \frac{1}{3\epsilon\sqrt{2\lambda}}. \qquad (8.21)$$

A fully symmetric solution with the big rips symmetrically arising around $t = 0$ can immediately be obtained by choosing $t_1 = 0$

The energy density and the pressure will be:

$$\rho = -\frac{2\lambda}{1 - 18\lambda\epsilon^2 (t - t_1)^2} < 0, \qquad (8.22)$$

$$p = \frac{2\lambda(1 + 2\epsilon)}{1 - 18\lambda\epsilon^2 (t - t_1)^2} > 0. \qquad (8.23)$$

Now, lets consider the universe filled with a scalar field ϕ, such that

$$\rho = K + V, \quad p = K - V,$$

with $K = \dot{\phi}^2/2$, $V = V(\phi)$. Using (8.22) and (8.23) one gets

$$K = \frac{2\lambda\epsilon}{1 - 18\lambda\epsilon^2 (t - t_1)^2} > 0, \qquad (8.24)$$

and

$$V = -\frac{2\lambda(\epsilon + 1)}{1 - 18\lambda\epsilon^2 (t - t_1)^2} < 0. \qquad (8.25)$$

Thus, in the brane world we have the highly nontrivial situation in which the model with a negative potential (and positive K) results in a big rip singularity.

One can see that

$$p + \rho = \frac{4\lambda\epsilon}{1 - 18\lambda\epsilon^2 (t - t_1)^2} > 0,$$

and

$$3p + \rho = \frac{4\lambda(1 + 3\epsilon)}{1 - 18\lambda\epsilon^2 (t - t_1)^2} > 0.$$

But since $\rho < 0$ both weak and strong energy conditions are violated. It is interesting to note that

$$\frac{\ddot{a}}{a} = \frac{6\epsilon\lambda \left(6\lambda\epsilon(t - t_1)^2 + 18\lambda\epsilon^2(t - t_1)^2 + 1\right)}{1 - 18\lambda\epsilon^2 (t - t_1)^2} > 0.$$

Finally, lets establish the exact functional dependence of V on the scalar field ϕ. To do it one needs to find $\phi = \phi(t)$ from (8.24), express $t = t(\phi)$ and then substitute it into (8.25). In our case we get

$$\phi(t) = \phi_0 \pm \frac{1}{3}\sqrt{\frac{2}{\epsilon}}\arcsin\left(3\epsilon\sqrt{2\lambda}(t - t_1)\right),$$

and hence

$$V(\phi) = -\frac{2\lambda(\epsilon + 1)}{\cos^2\left(3\sqrt{\frac{\epsilon}{2}}(\phi - \phi_0)\right)}, \tag{8.26}$$

where $\phi_0 = \text{const}$. Note that here the field ϕ is real-valued in spite of producing an essentially phantom dynamics.

Thus, we have a model with a positive tension, positive K, negative potential V, negative $\rho = K + V$ and positive pressure $p = K - V$. At the same time, $\rho + p > 0$ ($\rho + 3p > 0$ too) and $\ddot{a}/a > 0$. The universe appears to be filled with a scalar field with a rather amusing potential (8.26).

We have two big rips at $t = t_\pm$ (see Eq. (8.21)). In the "classical" limit $\lambda \to \infty$ we have only one big rip at $t = t_1$. Therefore, the situation with two big rips (initial and final) should be treated as an essential consequence of the existence of a brane.

So, what do we end up with?.. In short, a model which is a mirror opposite of a standard phantom cosmology. Where the phantom models have a negative pressure but a positive density, the Solution II predicts a negative density and a positive pressure (bounded from below by the condition $p > |\rho|$). This duality in signatures is further reflected by the behaviour of the scalar field potential (negative for the Solution II) and a kinetic term (which in this case is positive). As a result, we can call the Solution II a *dual phantom model*. Naturally, the one thing that the phantom and dual phantom scenarios have in common is the emergence of the future (and/or past) singularities.

If we look closer, we will see that the Solution II actually predicts two instances where the standard phantom energy (with negative kinetic term) becomes dominant: it happens in the time periods situated before the first big rip at t_- and after the second big rip at t_+ (see Fig. 8.3). Had we taken $w > -1$ in the precedent calculation then the resulting scale factor would have described a universe that contracts to a singularity (a big crunch from the point of view of an internal observer or a *big bang* with respect to the observer from the following region) at $a = 0$, both for a positive and a negative time, followed by two branches (one for positive t, one for negative) of the accelerated expansion extending to infinity. It is worth noticing that in this case, whereas the exterior accelerating regions are filled with a conventional dark energy, in the region sandwiched between the two zeros the matter would be dark energy's *dual*, for it will have negative values for ρ and $\rho + p$ (Fig. 8.3). Another noteworthy fact is that all the *dual* regions disappear in the limit $\lambda \to \infty$, which corresponds to a classical Friedmann cosmology with no brane. Therefore the existence of both dual regions and of a second big rip appears to be endemic to the brane models.

At this point the reader might remark that all the talk about the big rip singularities and the accelerated expansion in a brane world model is somewhat dubious. Surely, such phenomena ought to have been prevalent in the early universe, where the quantum gravity effects were maximal — and lose all the significance later on?.. However, we remind here that, since in the present model the absolute value of the energy density is an increasing function of time squared, the influence of both the brane and the quantum effects is expected to become relevant at later times instead of at the early stages.

The brane's symmetric solution, on the other hand, has a rather surprising property which can help to create and maintain a *shortcut* for the *interstellar travel*. In fact, following Krasnikov [102] one can define this shortcut as follows. Let C be a timelike cylinder in the *Minkowski space* L^4, let M be a globally hyperbolic spacetime and let U be a subset of M. Then U will be a required shortcut if there exist an isometry $\kappa : (M - U) \to (L^4 - C)$ and two points p and q such that $p \preceq q$ (i.e. there is a future-directed time-like curve from p to q), and it does not hold for $\kappa(p)$ and $\kappa(q)$. The two particular examples of such shortcuts are the wormholes (akin to the one we are going to consider in the next section) and the so-called *Krasnikov tube* [103], [104]. In order for a shortcut to exist the *weak energy condition* (WEC) $T_{\mu\nu} t^\mu t^\nu \geq 0$ (where $T_{\mu\nu}$ is the *stress-energy tensor* and t^μ is an arbitrary timelike vector) must be violated. Now, even though the violation of

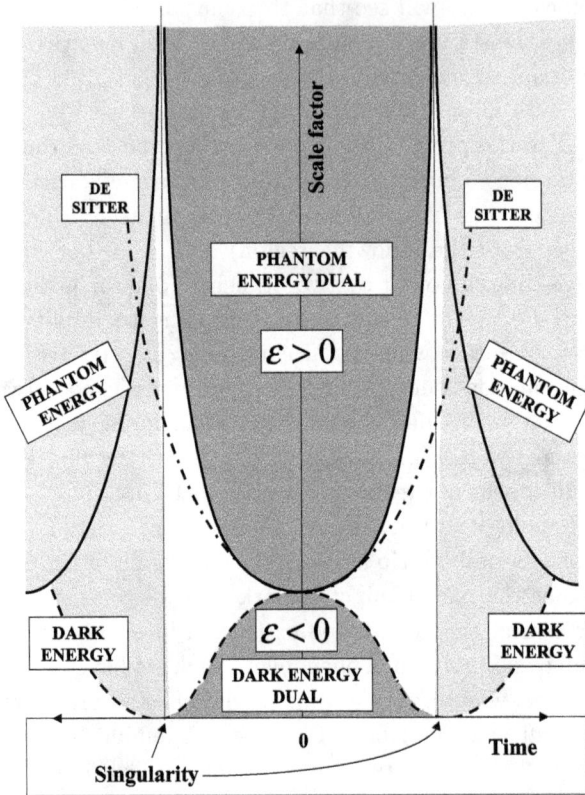

Fig. 8.3 The different dark energy and phantom energy regions which exhibited by Solution II. The bold line corresponds to the case $\epsilon > 0$, the dashed one — to the case $\epsilon < 0$, and the dash-and-dot is the De Sitter solution ($\epsilon = 0$). It can be seen that the dual regions are sandwiched between the two symmetric (big bang or big rip) singularities.

WEC in a quantum field theory is permissible (for example, in the *Casimir effect*) the *artificial creation* of the shortcuts (say, by a future technologically advanced civilization) is rather problematic because of the following reason. Ford and Roman ([105], [106]; see also [107], [108]) have shown that in the case of $d = 2$ massless scalar fields the following inequality holds

$$\rho_f \equiv \int_{\tau_1}^{\tau_2} d\tau \rho(\tau) f(\tau - \tau_0) \geq -|\tau_2 - \tau_1|^{-d}, \qquad (8.27)$$

where $\rho = T_{\hat{0}\hat{0}}$ ("hats" mean that one uses the orthonormal basis), d is the dimension of spacetime and f is a function such that $f \in C^\infty$, supp$f \in (\tau_2, \tau_1)$ and[2]

$$\int_{\tau_1}^{\tau_2} d\tau \frac{(f'(\tau))^2}{f(\tau)} \leq 1.$$

It is currently believed that this inequality shall apply for all cosmic solutions and shortcuts. In [102] (see also [104], [109]) it was actually shown that if the inequality (8.27) holds then one would need to have $E/c^2 = 3 \times 10^{62}$ kilograms of matter with the negative energy to construct a shortcut that would allow a transition of an object with a size of about 1 meter. Thus the condition (8.27) demonstrates that future manufacturers of the shortcuts would meet with serious, rather unsurmountable difficulties. However, if we consider solution (8.20) and take t_* as the absolute value of the time at which future big rip takes place, then using Eq. (8.22) one gets

$$\lim_{\tau_2 \to t_*} \int_{\tau_1}^{\tau_2} d\tau \rho(\tau) f(\tau - \tau_0) = -\infty.$$

Therefore the inequality (8.27) would no longer applies for a *RS brane* type I if it is filled with a phantom energy, so that the aforementioned problems with construction of the cosmic shortcuts are relaxed significantly.

8.3 Big trip and the gigantic black holes: a symmetric solution

Apart from their intrinsic interest, the cosmic solutions considered in Sec. 8.2 could in principle allow for the following possibility. It is known that as the universe moves towards the big rip singularity a very fast growth of the wormholes would commence. The growth will be so substantial that the diameter of an arbitrary wormhole's throat will become infinite well before the big rip is reached [95]. But the question is: would that still be true if one is to approach the big rip at a negative time? If this question is answered affirmatively then it would be only natural to assume that in the distant future a space-time bridge could be formed, reversibly linking the universe which is about to experience the future big rip with the same universe shortly *after* the past big rip. Or, in other words, we can then have a model which bears a striking similarity with the famous *Gödel*

---

[2]The additional condition $|\tau_2 - \tau_2| \leq \left(\max |R_{\hat{\mu}\hat{\nu}\hat{\rho}\hat{\sigma}}| \right)^{-1/2}$ must also hold.

model allowing for the *closed timelike curves* [110]. Moreover, if instead of a wormhole we would have a black hole swelling up into what we have dubbed a "big hole", then the bridge between the future and the past might also be formed, only this time in an irreversible fashion. In what follows we shall investigate this problem by considering the phantom energy accretion onto the wormholes and black holes in the framework of the cosmological solutions discussed in Sec. 8.2.

One can make a simple argument that would appear to preclude the existence of a symmetric couple of a swelled wormholes, linked to produces the big trip in the symmetric Solution I. This argument runs as follows. Similarly to an antiparticle being nothing but its counterpart moving backward in time, the exotic mass of a wormhole should be seen as just an ordinary matter in a Schwarzschild wormhole evolving with an external negative time. However, a wormhole with an ordinary matter is known to be unstable and pinches off immediately to convert itself into a black hole plus a *white hole*. Thus, a wormhole with a positive mass evolving backward in time is expected to be unstable. As one is going backward in time the first of the considered models becomes exactly equivalent to the phantom model moving forward in time due to a symmetric character of the solution, and therefore the positive energy density and the curvature both turn to infinity as t approaches $-t_R$. It follows that as one is approaching $-t_R$ there will be a phantom energy flow into the wormhole, decreasing its positive energy until it vanishes at the singularity at $-t_R$. This is the instability that prohibits a big trip to take place on the negative time branch of the solution. Whether or not one would accept this instability to be the same in case of the black holes in a quintessential phantom universe becomes a matter of interpretation. This heuristic prediction can actually be confirmed by direct calculations. In fact, if we write the Friedmann equations for a flat universe as

$$\left(\frac{\dot{a}}{a}\right)^2 = \rho, \tag{8.28}$$

$$2\frac{\ddot{a}}{a} = -(\rho + 3p), \tag{8.29}$$

we can rearrange them to produce

$$\frac{3}{2}(p + \rho) = -\frac{\ddot{a}}{a} + \left(\frac{\dot{a}}{a}\right)^2. \tag{8.30}$$

Now, the known expression for the rate of change of wormhole's mass due to the phantom energy accretion is [95]

$$\dot{m} = -\frac{3}{2}Am^2(p + \rho) \tag{8.31}$$

Integrating this equation with the use of Eqs. (8.30) and (8.31), we obtain

$$m = m_0 \left(1 - Am_0 \left(\frac{\dot{a}(t)}{a(t)} - \frac{\dot{a}(t_0)}{a(t_0)}\right)\right)^{-1} \tag{8.32}$$

We will use the following scale factor

$$a(t) = \alpha(\beta + \kappa t \cdot \tan(\kappa t)) \tag{8.33}$$

where α, κ and β are all positive constants. We shall study the solution (8.33) on the interval $-t_R < t < t_R$, with $t_R = \pi/(2\kappa)$. It is easy to see that $a(\pm t_R) = +\infty$, so that the Universe starts at a big rip singularity and thereafter undergoes a phase of contraction on $-t_R < 0 < t$ which ends at $t = 0$; then it starts expanding, finally ending again at a big rip singularity. With this scale factor we can obtain:

$$m(x) = m_0 \left[1 + Am_0H_0 - Am_0\kappa\frac{2x + \sin(2x)}{\beta\cos(x) + x\sin(2x)}\right]^{-1}, \tag{8.34}$$

where we have defined $x = \kappa t$, $-\pi/2 < x < \pi/2$. Then $m(x \to \pm\pi/2) = 0$. It is trivial to notice that $m(x) \neq m(-x)$, i.e., the growth of the wormhole does not preserve the symmetric character of the scale factor under the time reversal. As we shall see now this asymmetry leads to the emergence of a big trip (i.e. a blow-up of the wormhole throat size) on the positive-time branch, but not on the negative-time branch. In order to study the emergence of possible divergences in the case of Solution I, we subdivide the interval into three parts:

Subinterval $-\pi/2 < x < 0$ Here we can express the mass as

$$\frac{m(x)}{m_0} = [1 + Am_0(F(x) + H_0)]^{-1},$$

where

$$F(x) := -\kappa\frac{2x + \sin(2x)}{\beta\cos(x) + x\sin(2x)} > 0.$$

If there is a wormhole within the Universe, then it will always have a finite size on this subinterval.

Point x=0   At this point the wormhole throat will have a finite size which equals to $m(0) = m_0(1 + Am_0H_0)^{-1}$.

Subinterval $0 < x < \pi/2$ In this case, we have

$$\frac{m(x)}{m_0} = [1 + Am_0H_0 - Am_0G(x)]^{-1},$$

where

$$G(x) := -F(x) > 0.$$

It therefore follows that if a wormhole is to grow to infinite size, it must happen on this subinterval. The divergence points will be the zeros of the function $J(x)$, with

$$J(x) = (1 + Am_0H_0)\beta\cos(x) + [(1 + Am_0H_0)x - Am_0\kappa]\sin(2x) - 2Am_0\kappa x.$$

Here we see that $J(x)$ is continuous, $J(0) > 0$ and $J(\pi/2) < 0$; therefore there shall exist at least one zero on this subinterval.[3] That is, there is a big trip here but, as we have seen, there cannot be another one, symmetric to it and belonging to the corresponding negative time subinterval.

We note however that, even though at a negative time there would be no big trip to commence — a big hole would, even though we are dealing with a $w < -1$ model where no big hole can be expected. In order to see this, consider a sample black hole of mass M. The rate of change of M due to the accretion is given by [111]

$$\dot{M} = \frac{3}{2}BM^2(p + \rho),$$

with B being another numerical constant whose value is of the order of A. A straightforward calculation would then yield the following expression for the black hole mass

$$M(x) = M_0\left[1 - Bm_0H_0 + BM_0\kappa\frac{2x + \sin(2x)}{\beta\cos(x) + x\sin(2x)}\right]^{-1}. \quad (8.35)$$

Clearly, the zero of the denominator in this expression takes place on the subinterval at a negative time (the first of the subintervals we have discussed), but not at a positive one. This indicates that at a negative time a big hole phenomenon ought to take place. Unlike the quintessence models, such a phenomenon would have quite relevant effects in this case and

---

[3]Theoretically speaking, there can be more zeroes than just one, provided their total number is odd. However, because of the regular recurrence of the function $J(x)$ it is highly unlikely that there exist more than one zero on $0 < x < \pi/2$.

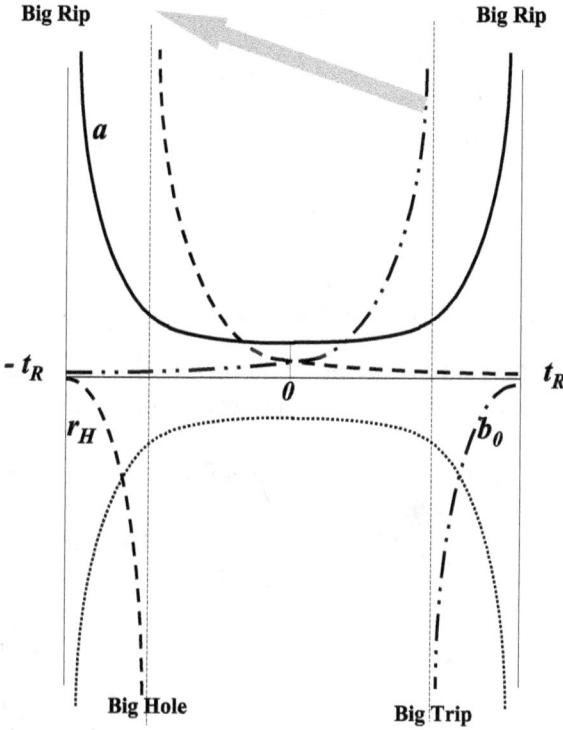

Fig. 8.4 Relative placements along time of the big trip and big hole processes that take place for Solution I. It can be observed that whereas a big trip of a wormhole occurs at a positive time, the big hole of black hole takes place at a negative time.

will be destined to occur for a reasonable set of parameters. In Fig. 8.4 we schematically show the processes of big trip and big hole for the Solution I.

Now the reader might ask: what would change in our big trip picture if we shall replace the Solution I with a Solution II, i.e. switch from a Friedmann to a brane scenario?.. Let's see!

The rate of change of the wormhole throat radius in the case of a brane symmetric solution ($t_1 = 0$) is

$$\dot{b} = 3A\epsilon\rho = \frac{6A\epsilon\lambda}{18\epsilon^2\lambda t^2 - 1}, \tag{8.36}$$

which, upon the integration, yields

$$b = \frac{b_0}{1 - \frac{A\sqrt{2\lambda}b_0}{2}\ln\left[\frac{(1-3\sqrt{2\lambda}\epsilon t)(1+3\sqrt{2\lambda}\epsilon t_0)}{(1+3\sqrt{2\lambda}\epsilon t)(1-3\sqrt{2\lambda}\epsilon t_0)}\right]}. \tag{8.37}$$

The zeros of the denominator of this function are

$$\tilde{t} = \frac{1-\xi}{1+\xi} t_*,$$ (8.38)

where $t_* = 1/(3\sqrt{2\lambda}\epsilon)$ is the absolute value of the time at which the big rips take place, and

$$\xi = \left(\frac{1 - \frac{t_0}{t_*}}{1 + \frac{t_0}{t_*}} \right) e^{2/(A\sqrt{2\lambda}b_0)}.$$ (8.39)

Therefore, we have the following four cases:

(1) If $0 < t_0 < t_*$, $\xi > 1$, then there will be a divergent wormhole swelling on $-t_* < t < 0$ (a situation that also corresponds to $0 < |t_0| < t_*$, with $\xi < 0$);
(2) if $0 < t_0 < t_*$, $0 < \xi < 1$, then there will be a divergent wormhole swelling on $t_* > t > 0$;
(3) if $t_0 > t_*$, $\xi < 0$ $|\xi| > 1$, then there will be a divergent wormhole swelling on $t < -t_*$ (a situation that also corresponds to $|t_0| > t_*$, with $\xi < -1$, $|\xi| > 1$);
(4) if $t_0 > t_*$, $\xi < 0$ $|\xi| < 1$, then there will be a divergent wormhole swelling on $t > t_*$.

What is common between all these distinct possibilities? Why, the fact that all of them predict a big trip (albeit at different times)!

A very similar situation is attained if we replace the wormhole with a black hole. In this case, the rate of change of the black hole mass M due to an accretion is

$$\dot{M} = -3B\epsilon\rho = -\frac{6B\epsilon\lambda}{18\epsilon^2\lambda t^2 - 1},$$ (8.40)

with which the following time-dependent black hole mass can be derived

$$M = \frac{M_0}{1 + \frac{B\sqrt{2\lambda}M_0}{2} \ln\left[\frac{(1-3\sqrt{2\lambda}\epsilon t)(1+3\sqrt{2\lambda}\epsilon t_0)}{(1+3\sqrt{2\lambda}\epsilon t)(1-3\sqrt{2\lambda}\epsilon t_0)} \right]}.$$ (8.41)

Thus, M will diverge (big hole) at exactly the times

$$\tilde{t} = \frac{1-\eta}{1+\eta} t_*,$$ (8.42)

with

$$\eta = \left(\frac{1 - \frac{t_0}{t_*}}{1 + \frac{t_0}{t_*}} \right) e^{-2/(B\sqrt{2\lambda}M_0)}.$$ (8.43)

The result of our endeavours then can be graphically represented as a sequence of all possible big holes distributed along the entire interval. The precise pattern of this distribution, along with a similar distribution of the big trips is displayed on Fig. 8.5. It will be discussed in the next section how all the regions of the complete time interval can be connected to each other without passing through the big rip singularities. Before we move

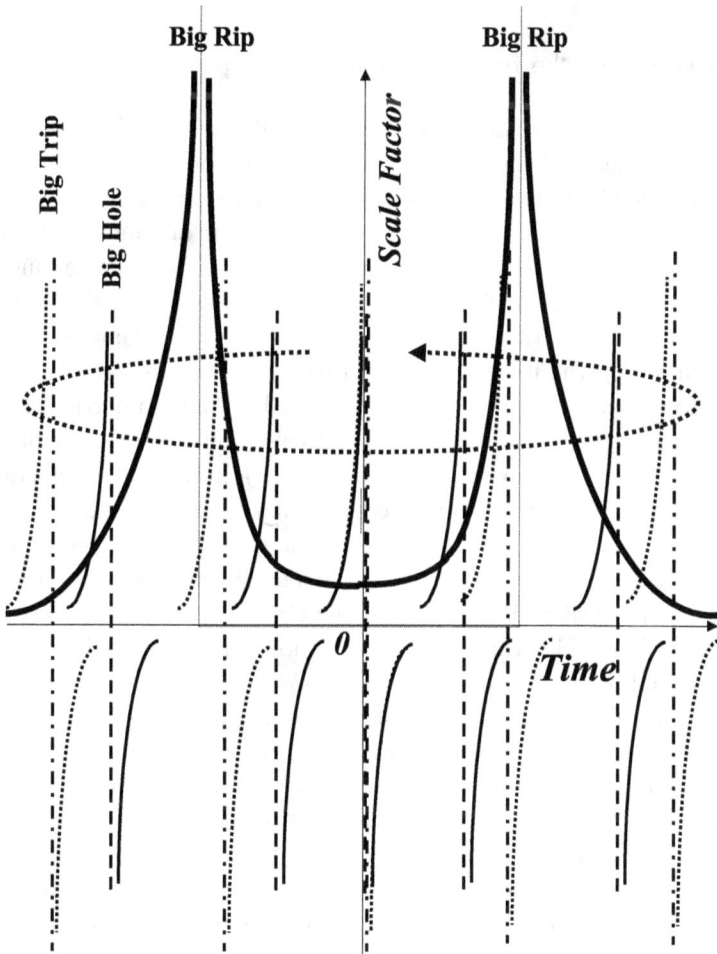

Fig. 8.5   The distribution of the big trip and big hole phenomena over time for Solution II. The exact placements will depend on the initial parameters characterizing the corresponding objects, b_0 for the wormholes and M_0 for the black holes, and on the time at which the observation is being made.

there, however, there is one further remark to make. It has been recently pointed out that, contrary to the classical Friedmann models, braneworld scenarios actually forbid the existence of static black holes [112]. Does it affect our considerations in any way?.. The answer is: not really. Since our calculations do not rely on the specifics of the black holes metric, as long as the brane cosmology allows for the formation of an object with an event horizon, the conclusion will stay the same.

8.4   Bridges to the future, bridges to the past

A *Morris-Thorne wormhole* can be converted into a *time machine* that allows any object passing through it to time travel into the past or the future, provided the two mouths of the wormhole are moving relative to each other [113]. Thus, since the space-time where the mouths of such a wormhole are located is itself dynamically expanding (or contracting), one should expect that a swelling wormhole might behave like a time machine. In other words, once the wormhole becomes infinite in diameter and literally swallows up the universe and everything in it, all its constituents — the entire universe — will then be transported in time. In particular, it has been shown in [95] that during the wormhole's swelling due to the phantom energy accretion the *chronology protection conjecture* [114] is violated, so that the macroscopic wormholes become quantum-mechanically stable during that accretion process. According to the distinct processes considered in the previous section we can have different kinds of bridges connecting the past and future of the universe, either circumventing a singularity type big rip, big bang, big crunch or failing to do so. In order to determine the structure and properties of these bridges, we have to take into account two requirements: (1) Any of the space-time swelling processes we have considered in this paper must be described for an asymptotic observer at radial coordinate $r \to \infty$, for otherwise such processes simply cannot take place or would lead to quite different behaviour [95], and (2) by their very definition, the spacetime of both a *Morris-Thorne wormhole* and a black hole ought to be asymptotically flat. This condition makes it impossible for the universe to travel along its own time through a single wormhole. We thus distinguish the following processes.

Single wormhole processes. While still smaller than the host universe, the swelling wormhole may allow for some given amounts of matter to time travel. Once the wormhole has grown up larger than the cosmological

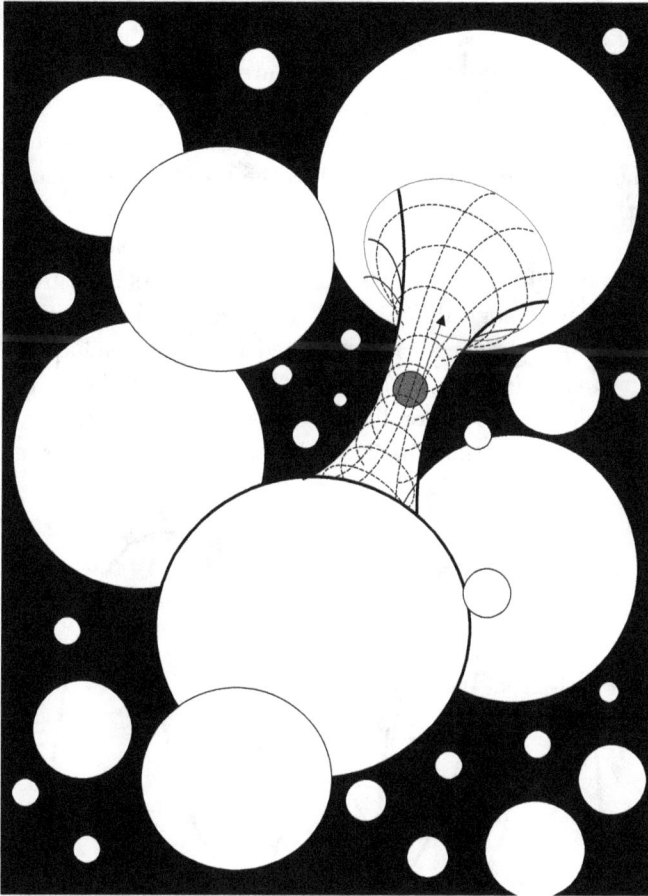

Fig. 8.6 Pictorial representation of the big trip process when it is carried out by a single grown-up wormhole within the framework of a multiverse picture. In this case the universe does not travel along its own time but behaves as if its whole content was transferred from one different large universe to another, larger universe.

horizon it can no longer insert its mouths into the host universe and, in order to become implanted anywhere, it must necessarily make a recourse to the different, sufficiently larger universes, if they are available at all (for example, in a multiverse scenario (see Fig. 8.6)), while satisfying the *Israel junction conditions*. The lack of a common time for the assumed set of universes would convert the big trip into a simple energy transfer process without any violation of causality. There is thus no proper time travel in the latter case.

Processes induced by a couple of swelling wormholes. As a couple of wormholes, one in the past and the other in the future, are growing bigger than the universe in which they are implanted, these wormholes should cease to insert their mouths onto the large regions of the universe where they were originally inserted. The resulting open mouths of one of the wormholes can be then connected to the resulting open mouths of the other, so that the two wormholes turn out to be mutually connected to each other in such a way that they form up a compact, closed quickly inflating tunnel right when their corresponding throats have grown beyond the size of the cosmological horizon. The universe trapped inside can then flow along the resulting traversable closed tunnel, travelling in this way along its own time (see Fig. 8.7). This "big kiss" can always be made to satisfy the Israel junction conditions and therefore effectively allows for the existence of connections between the past and future of the universe.

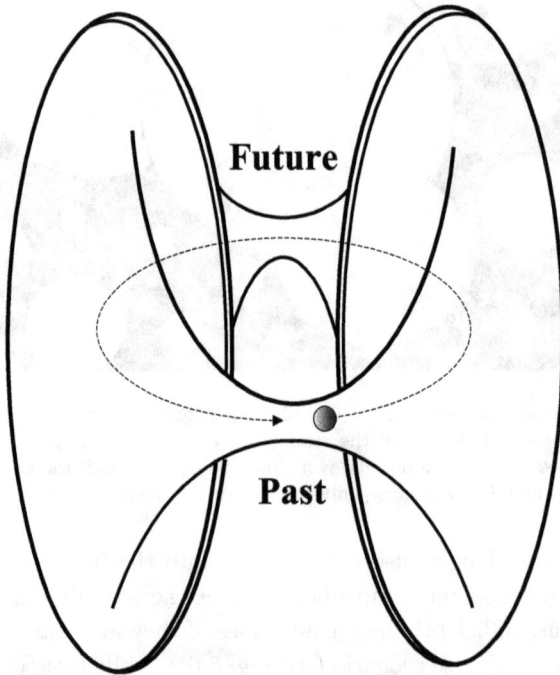

Fig. 8.7 Pictorial representation of the big trip process when two grown-up wormholes are used, one in the past and the other in the future. In this case the two wormholes connect their mouths in such a way that a compact tunnel is formed through which the universe can travel along its own time, into the past or the future.

Once the wormholes are annihilated, which would happen right after the big trip by converting themselves into a couple of black/white hole pairs, the universe would continue its conventional causal evolution, re-starting at the moment it has finished its time travelling.

Processes induced by the combined action of a swelling wormhole and a swelling black hole. The topology resulting in this case is like that of the two swelling wormholes considered in the precedent situation, except for one of the two wormholes being replaced with a swelling black hole. The effect is the same, too: the universe can time travel along its own time, in this case by using a black hole in the past or future as an intermediate stage on which the wormhole is inserted in a way that also satisfies the junction conditions. At first sight, it could be expected that the black hole would rapidly accrete all the exotic energy that keeps the wormhole throat open, thus making impossible our big trip scenario. That does not happen, however, due to the fact that the black hole swelling takes place only on the time branch (negative or positive) where its surface gravity becomes repulsive relative to an observer moving (forward or backward, respectively) in time.

Thus, the answer to the question posed at the beginning of section 8.3 is affirmative. Bridges linking the past and future of the universe may indeed be formed to mark the itineraries of closed timelike curves in a way fully reminiscent of the dream that Gödel had for most of his later life. Moreover, the wormholes by themselves may serve as a way of escaping from the initial and the ripping singularities. However, the question of what happens with the causality should still be addressed in such a scenario. In fact, the connection that would be established between the past and the future of the universe may lead to the *time travel paradoxes (TTP)*; that is, you for example can kill your parents before you were born (grandfather paradox) or you can kill yourself in the past.[4] Could paradoxes like these be avoided in our scenario? There actually are many articles dealing with this matter but the solution is still unclear. For example, Krasnikov [115] has defined the time travel paradoxes in physical terms, concluding that no paradoxes arise in general relativity. Another possible way to solve TTP is connected with the *many worlds interpretation of quantum mechanics* [116], but this solution is marred with its own problems [117].

---

[4] After all, this seems more humane than killing one's own parents!

A very attractive solution to the problem of TTP was suggested in the article [118]. For the case of a "hardsphere" self-interaction potential and the wormhole-based time machines, it was shown that for a particle with fixed initial and final positions traversing through the wormhole just once, the only possible trajectories that minimize the classical action would be those which are globally self-consistent. The principle of self-consistency (originally introduced by Novikov) becomes thus a natural consequence of the principle of minimal action. Although the verdict is still not settled down on the general validity of this solution (see, for example, [117]), we shall assume it to be correct. Then it will be shown in what follows that the flatness problem can be solved without making recourse to the inflationary paradigm. In order to understand how this is possible, let us consider the one-particle *Schrödinger equation*

$$i\hbar\frac{\partial\Psi}{\partial t} = -\frac{\hbar^2}{2m}\nabla^2\Psi + V\Psi. \tag{8.44}$$

Using Bohm substitution [119] $\Psi = Re^{iS/\hbar}$ (with real functions R and S) we can obtain two equations. One of them will be the Hamilton-Jacobi (H-J) equation for a single particle moving in a quantum potential

$$U = V - \frac{\hbar^2}{2m}\frac{\nabla^2 R}{R}.$$

The classical *principle of minimal action* will hold if $U = V$. In fact, classical mechanics assumes that a particle follows a path between the two points that minimizes the classical action S. In quantum theory all possible paths contribute to the path integral. Thus the principle of minimal action will hold in the present case if $\nabla^2 R = 0$. On the other hand, the latter equation has no bounded nonsingular solutions unless we have $R = $ const.[5] The simplest example is a plane wave solution in nonrelativistic quantum theory, with a wave function that describes a free particle.

The same idea can be extended for the quantum cosmology and the *Wheeler-DeWitt equation* $\hat{H}\psi = 0$, where $\hat{H}$ is the super-Hamiltonian operator and ψ is the wave function of the universe. As a result we will have $\psi = Re^{iS/\hbar}$, where $R = $ const. In other words, R does not depend on the scale factor a. Such non-normalizable wave function of the universe was originally suggested by Tipler in [120] by introducing as a boundary

---

[5] Here we shall replace the Hilbert space with a more general rigged Hilbert space which includes the delta functions.

condition the requirement that the quantum state of the universe should allow Einstein equations to hold exactly in the present epoch. One can therefore conclude that this boundary condition must be satisfied in order to avoid the emergence of TTP in cosmology, provided the conclusion in [118] is correct. We note however that in the neighborhood of the big rip singularities Einstein equations cannot hold and therefore TTP could then take place.

Moreover, by using a wave function with $R = $ const Tipler also showed how the *flatness problem* can be solved. To see how, let us consider the probability that we will find ourselves in a closed universe with the radius larger than any given radius a_{given}

$$\int_{a_{given}}^{+\infty} |\psi|^2 da = +\infty,$$

whereas

$$\int_{0}^{a_{given}} |\psi|^2 da < +\infty,$$

so the *relative* probability that we will find ourselves in a universe whose radius would be larger than any given radius a_{given} is equal to one. Thus, using the condition that *there are no TTP* one can also solve the flatness problem in cosmology.

To make this point clearer let us return to the case of (8.44). If $V = 0$ then the solution of (8.44) has the form:

$$\Psi = e^{-i\left(\hbar \vec{k}^2 t/(2m) + \vec{k}\vec{r}\right)}. \tag{8.45}$$

In the lab we have some prior information about an initial location of the particle so one must utilize a wave packet rather then plane wave (8.45). But if we do not have any (prior) information then we are compelled to use the non-normalizable wave function (8.45). In this case we cannot calculate the probability $p(V < V_{given})$ to find the particle inside the given volume V_{given} and one can conclude that $p(V < V_{given}) \sim V_{given}$. The probability to find this particle out of this volume but inside the volume V_1 will be proportional to $V_1 - V_{given}$. Now a *relative* probability can be obtained exactly:

$$p_{rel}(V < V_{given}) = \frac{V_{given}}{V_1 - V_{given}}.$$

It follows that the relative probability to find the particle inside of the volume V_{given} ($V_1 \to +\infty$) will be zero and the relative probability to find this particle outside this volume will be one. *This is a direct consequence of having no prior information.*

This situation is unusual for the lab but natural for the quantum cosmology. In the latter case we do not have any prior information and one has to use the non-normalizable wave function of the universe ψ instead of, say, a wave packet. It is well known that the wave function of the universe is non-normalizable in a tree-level approximation and this problem can be solved by including the loop effects. But in the framework of our approach, the wave function in a tree-level approximation becomes the true wave function and all loop effects must be suppressed. Our conclusion therefore is that when avoiding time travel paradoxes the way we have outlined above, one gains an unexpected reward: the solution of the flatness problem that does not require a recourse to the inflationary ideas.

8.5   Remarks and reminiscences

In the last few sections we have looked at a number of symmetric cosmological solutions for the late or early universe by using some special forms of dark energy that either satisfy the energy conditions or violate them in several different ways. The distinctive property of all such solutions is that they can double in a symmetric manner the main singular events predicted by the corresponding quintessential cosmologies. In particular, they clone the big rip and the big bang (or big crunch) and make them appear twice, one on the positive branch of time and the other on the negative one. However, the effects of the accretion of this generalized dark energy onto black holes and wormholes do not display such a symmetry so that the big trip only appears once, either on the positive branch or on the negative branch of time. An interesting aspect of our investigation touched on the possibility of making viable a relevant swelling of black hole space-times so that it may lead to the so-called "big hole" phenomenon by which the black hole grows big enough to encompass (swallow) its entire parent universe. That big hole was precluded in quintessence models and is also predicted to appear here just once, either on positive or negative time.

The most interesting aspects of the above solutions were derived from the brane world scenario, as it has been shown that the prediction of a second singular events is an entirely brane-reliant effect and that the late evolution of the universe predicted within such brane models is due to the existence of an energy density whose absolute value increases with cosmological time, thus making the brane and quantum effects to grow and become dominant mainly at the later stages of the cosmic evolution. On the other hand, the simultaneous emergence of the big trip and big hole phenomena before and after the cosmic singularities makes it possible to circumvent such singularities so that the full interval for the universe evolution is extended to cover the entire range from $t = -\infty$ to $t = +\infty$.

It should also be remarked that in case that $\epsilon < 0$ in the symmetric solution derived from the brane model (Solution II), the two symmetric zeros of the scale factor at $t_{\pm} = \pm t_{bb} = 1/(3\sqrt{2\lambda}|\epsilon|)$ actually correspond to two big bang (or big crunch) singularities where the energy density diverges. Such singularities can be also circumvented as, like in the case of a positive ϵ, the wormhole's and black hole's swelling and the corresponding big trip and big hole, will take place at

$$t = T = \frac{\xi - 1}{\xi + 1} t_{bb}, \tag{8.46}$$

with

$$\xi = \left(\frac{1 + \frac{t_0}{t_{bb}}}{1 - \frac{t_0}{t_{bb}}} \right) e^{2/(A\sqrt{2\lambda}b_0)} \tag{8.47}$$

for wormholes and

$$\xi = \left(\frac{1 + \frac{t_0}{t_{bb}}}{1 - \frac{t_0}{t_{bb}}} \right) e^{-2/(B\sqrt{2\lambda}M_0)} \tag{8.48}$$

for black holes, and would again crop up along the entire time interval from $-\infty$ to $+\infty$ in such a way that, relative to different observers placed on distinct regions of that interval, there will be no need to pass through the big bang (or big crunch) singularities. This mechanism may provide us with a new alternative for a smooth creation of the universe, other than the *Hawking no-boundary condition* [121] or the *Gott's noncausal self-creating condition* [122].

Before closing up, a quite interesting point is worth mentioning. The checked possibility of establishing causality-violating links between the region inside the two big rips and the regions outside that interval for brane solutions (8.20) makes it unavoidable that the space-time to be considered extends from $-\infty$ to $+\infty$ without passing through the big rip singularities. However, for this to be possible it is necessary that the scale factor be well-defined along this infinite interval, a case which can only be satisfied if the parameter ϵ (or $|\epsilon|$ in case of the solution containing two symmetric big bangs) entering the equation of state is discretized so that [123]

$$\epsilon = \frac{1}{12(n+1)}, \quad n = 0, 1, 2, 3, \ldots, \infty. \tag{8.49}$$

Even though we do not quite understand yet the deep physical meaning of this requirement one can still say that: (i) it leads to a preliminary "quantization" of both the involved dynamical quantities such as the energy density, pressure, potential energy and scalar field, and the space-time quantities such as the occurrence time for big bang and crunch, big rip, big trip and big hole, and (ii) it makes any future or past event horizons to disappear, allowing for any amount of information to be transferred during the big trip process, and for the formulation of the fundamental theories based on the definition of an S-matrix (such as string or M-theories), to be mathematically consistent. It is tempting to speculate nevertheless that the discretization of the equation of state parameter could be regarded to be at qualitatively the same footing with respect to a proper quantum theory of the universe as the original Bohr theory did in relation with the final probabilistic description of the quantum mechanics of a hydrogen atom.

Chapter 9

Phantom multiverse: a curious synergy of cosmology and particle physics

9.1 A dark energy multiverse

The perception that the very big and the very small are both governed by the same physical laws is an ancient concept that has rendered extremely fruitful results such as the Galileo's mechanics or the Newton's universal gravitation law. That concept is actually a very popular one that is being adopted by many people at all cultural levels. However, the current status of particle physics and cosmology seems to inexorably thwart such a common treatment. In fact, by the time being, in spite of the success of the inflationary paradigm, they look irreconcilable indeed [124]. There are several main points for that discrepancy. In particular, the known *problem of the cosmological constant* [125] by which the predicted value of the quantum-field vacuum energy density and that for cosmology are currently separated by many orders of magnitude, the feature that whereas fundamental physical theories are time-symmetric, cosmology contains an intrinsic *arrow of time* [126], and the existence of a *future event horizon* with finite proper size in accelerating cosmology which is mathematically and physically incompatible with any fundamental theory, such as string theory or quantum gravity, based on the introduction of an S-matrix requiring the propagation between points infinitely space-separated [127]. Actually, that can be regarded to be one of the greatest problems of all theoretical physics.

We take the above difficulties as being fundamental and essentially inescapable, provided that one views oneself, the physics of the particles and the cosmology all lying within the realm of one unique single universe like the one which we live in. Really, what we are going to argue is that all these difficulties simply vanish provided we adopt a multiversal view on the universe, in which the proper fundamental laws of physics shall be formulated for the multiverse composed of the universes that exist for *finite* periods of time. In contrast, our observed cosmology describes merely one universe, endowed with an infinite cosmic time (not identical to the multiversal time!), isolated from the multiverse but with the physical characteristics being precisely and consistently relatable to those of the entire multiverse.

There already are a plethora of *multiverse models*, essentially including those coming from quantum mechanics [128], those described in inflationary theory [84], those which come about in string or M theories [129], and those which are just based on classical general relativity [130]. In order to implement our main idea we should choose a classical multiverse model that is able to account for the current accelerating cosmic behaviour, leaving all quantum considerations to be built up later on. One of such multiverse models is the recently suggested dark energy multiverse [131]. It is the goal for this chapter to discuss and understand the possible ramifications of such model.

This idea of the multiverse and its relationship with the physics of a particular universe in it reminds us of the *Plato's cave* [132]. In this myth a group of humans is chained up in a cave with their backs turned to the entrance and a bonfire raging right between them and this entrance. As a result, the humans are condemned to see only the shadows of the world outside. So Plato explains the relation between the observable world, which we actually perceive, and the world of ideas, the real world. What we are planning to do in this chapter is to attune this philosophical concept to a physical reality, by noting that it is the true physical nature that serves the role of an exterior of a cave, which we can only measure (perceive) by observing its "shadow". The real world in question would be the entire multiverse, where the particle physics is well defined, while the world that we can experience, the cosmological world of Saturn, Tau Ceti and Andromeda Galaxy, will be but a shadow, a projection from the real world. Then the apparent inconsistencies that arises when we compare the particle physics and the cosmology can be seen as mere artifacts arising from trying to identify the shadows with the real world.

We shall now briefly review the dark energy multiverse, a scenario based on accelerating cosmology which has been recently introduced in [131]. If we consider a cosmological model which is dominated by a quintessential fluid with constant equation of state $p = w\rho = w\rho_0[a(t)/a_0]^{-3(1+w)}$ (where $a(t)$ is the scale factor for the universe, and p and ρ are respectively the pressure and energy density, with the subscript "0" denoting current value) plus a negative cosmological constant, Λ, then the Friedmann equation can be written as

$$H^2 = -\lambda + Ca^{-3\beta}, \qquad (9.1)$$

where $\lambda = |\Lambda|/3$, $C = 8\pi\rho_0/(3a_0^{-3\beta})$ and $\beta = 1+w$. By integrating Eq. (9.1), we can obtain the time evolution of the cosmic scale factor, for $\beta < 0$,

$$a(t) = a_0 \left[\cos\left(\alpha(t - t_0)\right) - b\sin\left(\alpha(t - t_0)\right)\right]^{-\frac{2}{3|\beta|}}, \qquad (9.2)$$

with $\alpha = \frac{3|\beta|}{2}\lambda^{1/2}$ and $b = \left(\frac{C}{\lambda}a_0^{-3\beta} - 1\right)^{1/2} = \left(\frac{8\pi}{3\lambda}\rho_0 - 1\right)^{1/2}$. Therefore, an infinite number of big rip-like singularities will occur in this model at times given by

$$t_{br_m} = t_0 + \frac{2}{3|\beta|\lambda^{1/2}}\text{arctg}\left[\left(\frac{8\pi\rho_0}{3\lambda} - 1\right)^{-1/2}\right] + \frac{2m\pi}{3|\beta|\lambda^{1/2}}, \qquad (9.3)$$

in which m is any natural number. It is easy to check that, for $m = 0$, expanding the above expression for $\lambda \ll 1$, we recover the occurrence time of the big rip for a quintessence model of phantom energy without any cosmological constant, that is,

$$t_{br} = t_0 + \frac{1}{|\beta|\,(6\pi\rho_0)^{1/2}}. \qquad (9.4)$$

Even though we have used a negative cosmological constant, the fact that one can define an overall positive vacuum energy density has led us to a true cosmic model. It is worth noticing that in order to make all possible physical regions in the above solution physically meaningful, the parameter of the equation of state should be discretized, $|\beta| = \frac{1}{3n}$, with $n = 1, 2, 3, \ldots$, so guaranteeing the scale factor to be always positive.

Due to the singular character of the big rips, in the absence of any wormhole-type connection between single universes, the regions between two such singularities are causally disconnected and each of these regions can be interpreted as a different spacetime (see [131]), in fact a different universe within the whole infinite multiverse. Classically and in the absence of observable matter, all of these universes in the multiverse are physically indistinguishable; all starting with an infinite size, which will then steadily decrease until a minimum nonzero value,

$$a_{\min} = a_0 \left(\frac{8\pi\rho_0}{3\lambda}\right)^{-1/3|\beta|} > 0. \qquad (9.5)$$

After that moment the universe acceleratingly expands again to infinity.

We finally want to remark that even though all the individual universes in the multiverse are classically identical, it could well be that one might envisage the contents of the set of universes acting as a physical realizations of the quantum superposition in a quantum-mechanical treatment, somehow parallelling the *Everett's many-world interpretation*.

It is also worth noticing that a multiverse scenario with essentially the same structure as the one discussed above might be produced in a framework of a *Randall-Sundrum brane* with a positive tension, $\mu > 0$, and a negative cosmological constant. In fact, there are different models stemming from this scenario depending on the features of this cosmological constant, since it can exist inside of a bulk (then the observable cosmological constant is but a projection), or on the brane. In the first case, the bulk is filled with phantom energy with equation of state $p = w\rho$ ($w = -1 - \alpha/3$, $\alpha > 0$) and a negative cosmological constant, $\Lambda < 0$, such that $\lambda = -\Lambda/3 > 0$. Then the scale factor evolution in the brane is governed by the modified Friedmann equation [101],

$$H^2 = \rho_T \left(1 + \frac{\rho_T}{2\mu} \right), \tag{9.6}$$

where $\rho_T = -\lambda + Da^\alpha$, with D being a constant. Integrating Eq. (9.6) for $2\mu > \lambda$ and appropiatelly choosing the integration constant, one obtains

$$a(t)^\alpha = \frac{\lambda(2\mu - \lambda)}{D \left[2\mu \cos^2 \left(\frac{\alpha\sqrt{\lambda(2\mu-\lambda)}}{2\sqrt{2\mu}} t \right) - \lambda \right]}. \tag{9.7}$$

It must be noticed that, in order to have a well defined scale factor in the brane a discretization is needed: $\alpha = 1/(2n)$, with $n = 1, 2, \ldots$. Now we can immediately see that there will be an infinite number of big rip singularities at

$$t_{br_m} = \frac{2\sqrt{2\mu}}{\alpha\sqrt{\lambda(2\lambda - \lambda)}} \left(\pm \arccos\sqrt{\frac{\lambda}{2\mu}} + 2\pi m \right), \tag{9.8}$$

or

$$t_{br_m} = \frac{2\sqrt{2\mu}}{\alpha\sqrt{\lambda(2\lambda - \lambda)}} \left(\mp \arccos\sqrt{\frac{\lambda}{2\mu}} + 2\pi(2m \pm 1) \right). \tag{9.9}$$

Thus, one can construct a multiversal scenario fully analogous to the one discussed above for brane worlds with $\mu > \lambda/2$. This will be no longer the case however if $\lambda \geq 2\mu$. In fact, when $\lambda > 2\mu$, we get the solution

$$a(t)^\alpha = \frac{\lambda(\lambda - 2\mu)}{D\left[2\mu \sinh^2\left(\frac{\alpha\sqrt{\lambda(\lambda-2\mu)}}{2\sqrt{2\mu}}t\right) + \lambda\right]}, \tag{9.10}$$

which does not show any big rip singularities and therefore cannot be cut off in an infinite set of independent universes, and if $\lambda = 2\mu$ such a solution reduces to

$$a(t)^\alpha = \frac{4\lambda}{D\left(4 - \lambda\alpha^2 t^2\right)} \tag{9.11}$$

that describes a single universe which starts and dies at big rip singularities taking place at

$$t_{br} = \pm\frac{2}{\alpha\sqrt{\lambda}}. \tag{9.12}$$

Thus, the last two subcases could not lead to any multiverse scenarios. As it is mentioned above, there exists another procedure to take into account a negative cosmological constant in a *Randall-Sundrum scenario* this time adding the cosmological-constant term directly to the modified Friedmann equation, i.e., assuming the cosmological constant to be defined on the brane

$$H^2 = \rho\left(1 + \frac{\rho}{2\mu}\right) - \lambda, \tag{9.13}$$

where $\rho = Da^\alpha$ is the phantom energy density in the bulk. Integrating Eq. (9.13) one obtain

$$a(t)^\alpha = \frac{2\lambda}{D\left[1 - \left(\frac{\mu+2\lambda}{\mu}\right)^{1/2}\sin(\alpha\lambda^{1/2}t)\right]}. \tag{9.14}$$

So, once again, the spectrum for the equation of state parameter α must be discrete, $\alpha = 1/(2n)$ with $n = 1, 2, \ldots$. The scale factor diverges infinitely many times at points

$$t_{br_m} = \frac{1}{\alpha\lambda^{1/2}}\arcsin\left(\frac{\mu}{\mu + 2\lambda}\right)^{1/2} + \frac{2m\pi}{\alpha\lambda^{1/2}}, \tag{9.15}$$

which takes place for all values of $\mu > 0$ and $\lambda > 0$, therefore there is a multiverse scenario independently of these values.

In the rest of this chapter we shall restrict ourselves to the braneless multiverse case, leaving the treatment of the brane multiverse to be dealt with elsewhere. We only point here that a brane world topological defect must split into an infinite set of individual defects out from which just one may develop the physical properties that we can observe in our universe.

9.2   A single observable universe?

Singling out a universe from the whole multiverse should imply the consideration of observers in that single universe that would in principle interpret their spacetime as the unique, full spacetime for which, in the absence of past or future singularities or crunches, the allotted time ought to stretch to infinity. Since every single universe in the multiverse is defined for a finite time interval, our first task must be to re-scale the finite time interval of one such single spacetimes so that it becomes infinite.

Thus, out from Eq. (9.2) we first consider a single finite-time universe whose scale factor is expressed as

$$a(\tau) = a_{\min} \cos^{-\frac{2}{3|\beta|}}(\tau), \qquad (9.16)$$

where $-\pi/2 \leq \tau \leq \pi/2$ with $\tau = \frac{3|\beta|}{2}\lambda^{1/2}(t - t_0) + \arctan\left(\frac{H_0}{\lambda^{1/2}}\right)$, so that $a(\tau)$ reaches its minimum value at $\tau = 0$. Next we re-define the time τ so that the new infinite time is given by $T = \tan(\tau)$, with $-\infty \leq T \leq \infty$. As expressed in terms of time T the scale factor would become

$$a(T) = \frac{a_{\min}}{\cos^{2/(3|\beta|)}(\arctan T)} = a_{\min}(1 + T^2)^{1/(3|\beta|)}. \qquad (9.17)$$

In order to obtain a more familiar expression for the scale factor, it is convenient to re-define again T in terms of another infinite time given by

$$\eta = \frac{1}{\lambda^{1/2}}\operatorname{arccosh}\left(1 + T^2\right), \qquad (9.18)$$

so that

$$a(\eta) = a_{\min}\cosh^{1/3|\beta|}\left(\lambda^{1/2}\eta\right), \qquad (9.19)$$

with $-\infty \leq \eta \leq \infty$ again.

Thus, we have been able to derive an expression for the scale factor which somehow resembles that for a de Sitter space. Even the Eq. (9.19) yields a scale factor compatible with the one of a de Sitter model, provided we consider a special case $n = 1$, i.e. $|\beta| = 1/3$. However, in order to see how our model adjusts to the available observational data, one need to use the current value of the hyperbolic cosine in Eq. (9.19), whose expression in terms of H_0 and λ can be derived by two distinct procedures: either by directly specializing Eq. (9.19) to $\eta = \eta_0$, without any need of differentiating $a(\eta)$ with respect to η, or by first using the definition of H in terms of the $\dot{a}(\eta) = da(\eta)/d\eta$ derivative and then specializing to $\eta = \eta_0$. Of course, for the physical model to be consistent the above two procedures should yield the same final expression. However, what we get instead is $\cosh\left(\lambda^{1/2}\eta_0\right) = (H_0^2 + \lambda)/\lambda$ following the first procedure and $\cosh\left(\lambda^{1/2}\eta_0\right) = \left(\frac{\lambda - (3|\beta|)^2 H_0^2}{\lambda}\right)^{-1/2}$, using the second one. The ultimate reason for such a discrepancy should reside indeed in the feature that the Friedmann equation (9.1) describes the whole multiverse, not every single isolated universe in it. Actually, the most general Friedmann equation which is compatible with a generic functional form for the scale factor like in Eq. (9.19) can be checked to be

$$H^2 = C_n a^{-3\beta_n} + \lambda_n, \tag{9.20}$$

in which $C_n = 8\pi\rho_{n0}a_0^{3\beta_n}/3 < 0$, $\beta_n = 1 + w_n > 0$ and $\lambda_n = \Lambda_n/3 > 0$, with ρ_{n0}, w_n and Λ_n being general values to be specified later on. Eq. (9.20) admits the solution

$$a(t_c) = a_{\min}\cosh^{\frac{2}{3\beta_n}}\left(\frac{3\beta_n}{2}\lambda_n^{1/2}t_c\right), \tag{9.21}$$

where $a_{\min} = a_0\left(\frac{8\pi|\rho_{n0}|}{3\lambda_n}\right)^{\frac{1}{3\beta_n}}$.

Equalizing finally Eqs. (9.19) and (9.21) we get

$$\beta_n = 2|\beta| = \frac{2}{3n}, \tag{9.22}$$

$$\left(\frac{8\pi|\rho_{n0}|}{3\lambda_n}\right)^{\frac{1}{3\beta_n}} = \left(\frac{8\pi\rho_0}{3\lambda}\right)^{-\frac{1}{3|\beta|}} \tag{9.23}$$

and

$$(\lambda)^{1/2}\eta = \frac{3\beta_n}{2}(\lambda_n)^{1/2}t_c. \tag{9.24}$$

From Eqs. (9.22) and (9.23), using the definition of λ and λ_n, we have

$$(8\pi)^3\frac{|\rho_{n0}|}{\Lambda_n} = \frac{|\Lambda|^2}{\rho_0^2}. \tag{9.25}$$

Now, the above two distinct ways to check the consistency of the model both produce the same expression for the hyperbolic cosinus as referred to current time t_{c0},

$$\cosh^2\left(\frac{3\beta_n}{2}\lambda^{1/2}t_{c0}\right) = \frac{\lambda}{\lambda - H_0^2}, \tag{9.26}$$

so implying that the new time t_c makes a well-defined choice for the cosmic time.

Besides the above argument, a full consistency of the model requires from it a suitable acceleration; more precisely, we need the expression for $\kappa = -q_0$ derived from the Friedmann equation for a flat geometry to be the same as the one obtained by directly applying the definition of q_0 in the present model. We also require the predicted value of κ to be compatible with the one which is expected for an accelerating universe, that is to say [133], κ should be slightly greater than unity.[1] In fact, if we have $\Omega_T = \Omega_n + \Omega_{\Lambda_n} = 1$, with $\Omega = \rho/\rho_{\text{crit}}$, $\rho_{\Lambda_n} = 3\Lambda_n/(8\pi)$ and $w_{\Lambda_n} = -1$, and from the second Friedmann equation

$$3\frac{\ddot{a}}{a} = -4\pi(3p_T + \rho_T), \tag{9.27}$$

we can get

$$\frac{\ddot{a}_0}{a_0} = H_0^2\left(\frac{3\beta}{2}|\Omega_n| + 1\right), \tag{9.28}$$

and whence

$$\kappa = \frac{3\beta}{2}|\Omega_n| + 1 > 1, \tag{9.29}$$

which should in fact be just slightly greater than unity as β and $|\Omega_n|$ are both very small as we will see later on.

Now, directly differentiating the scale factor given by Eq. (9.21) and using Eq. (9.26) we finally obtain

$$\kappa = \frac{3\beta}{2}\left(\frac{\lambda}{H_0^2} - 1\right) + 1, \tag{9.30}$$

which can readily be seen to be the same as (9.29).

Once we have checked the above consistency criteria, let us consider the physics of the resulting cosmological model for one single universe in the multiverse. Actually, after starting with a cosmic model equipped with a negative cosmological constant plus a vacuum phantom fluid characterized by a positive energy density and $\beta < 0$, so that the total vacuum energy density was positive, we have finally singled out an observable universe which still has a positive total vacuum energy density but now distributed

---

[1]The WMAP 3-year data, in combination with large-scale structure and SN data, allows for $w < -1$ in a general relativistic model. For constant w_0 [133] $w_0 = -1.06^{+.13}_{-.8}$ and $q_0 = (1 + 3w_0)/2$.

as a sum of a positive cosmological constant and a negative dynamical part. The latter part corresponds to the so-called dual of dark energy, or in short, *dual dark energy*. It is generally defined as a fluid having negative energy density and positive pressure, with β_n taking on values from 0 to 2/3. Besides the important property that the total energy density of the model is definite positive, such a negative dynamic density-energy component violates most energy conditions [134] and only may be allowed to exist provided that it is very small, in a quantum-mechanical context. This condition will be shown to be fulfilled in Sec. 9.3 and, since the quantum inequality condition [105] that any existing negative energy should be always accompanied by an overcompensating amount of positive energy (here given by the cosmological constant terms) is also satisfied, it appears that the resulting cosmic model fulfills all observational requirements and can be expected to provide us with a consistent and realistic scenario to deal with current cosmology. In fact, such a scenario is again somehow similar to a de Sitter framework and appears to also have a cosmological horizon. Because $a(t_c) = a_{\min}\cosh^n\left(\frac{3\beta_n}{2}\lambda_n^{1/2}t_c\right)$, where $n = 1, 2, 3, \ldots$ and the case for $n = 1$ corresponds to de Sitter universe, the acceleration predicted by these models goes generally beyond that of a de Sitter space, without giving rise to any future singularity of the big rip type, a case which is certainly compatible with nowadays observational data.

9.3 Is there a multiversal link between fundamental physics and cosmology?

At the beginning of Sec. 9.1 it was pointed out that in spite of belonging to a long and fruitful tradition the idea that the very large and the very small are both governed by essentially the same laws has reached a turning point during the last decades from which one only finds failures in its application to current particle physics and cosmology. Moreover, it is not just that such laws have some disagreements, but that these two branches of physics appear to be actually incompatible. In this chapter we are going to distinguish the three main situations where the discrepancies are most apparent: the so-called problem of the cosmological constant, the existence of an arrow of time in cosmology and the current cosmological prediction that there exists a future event horizon whose proper size is finite. In what follows we shall argue that the headaches produced by these three apparently basic difficulties all vanish in the multiverse framework. We are basically arguing that the above three shortcomings are nothing but artifacts coming from considering as the universe what really is nothing but a part of the whole physical reality. That is to say, whereas the fundamental physics resides and is well-defined in the whole realm of the multiverse (or just in one of its finite-time universes if all the multiversal components are identical), what we usually take as cosmology is defined just for one of the infinite independent spacetimes which the multiverse is made of when it is referred to a suitable infinite cosmic time.

In order to satisfy the observational data it is necessary that $H_0^2 \sim \lambda_n \sim 10^{-52} \mathrm{m}^{-2}$ [135]. Furthermore, the Friedmann equation for a infinitely prolonged individual universe, Eq. (9.20), imposes that $\lambda_n > H_0^2$ as C_n is negative. It follows that the absolute value of ρ_n must be very small and therefore negative energies could only appear in our observable universe when they are small enough. On the other hand, if we consider that the absolute value of the constant vacuum energy density in the multiverse corresponds to the Planck value, i.e. $\rho_\lambda^{Pl} \sim 10^{110} \mathrm{erg/cm}^3$, then $\lambda \sim 10^{62} \mathrm{m}^{-2}$. Taking into account relation (9.23) we then have

$$\rho_0 \sim \frac{10^{35} \mathrm{m}^{-3}}{|\rho_{n0}|^{1/2}}. \tag{9.31}$$

In addition, in order to have a positive definite Hubble parameter also in the multiverse it is required that $\rho_0 > 3\lambda/(8\pi) \sim 10^{61} \mathrm{m}^{-2}$. From these considerations, it follows that

$$|\rho_{n0}| \sim \frac{10^{70}}{\rho_0^2} < 10^{-52} \mathrm{m}^{-2}. \tag{9.32}$$

Because this should be always satisfied, as we have pointed out above, we can finally conclude that in the present scenario it is natural to have a cosmological constant in the multiverse with a value compatible with high energy physics and simultaneously a much smaller value for that constant of the order of those predicted in current cosmology in our observable universe.

Any cosmological model, including the one which gives rise to the multiverse, possesses an intrinsic *arrow of time*, that is a privileged direction along which the time flow is strictly unidirectional, from past to the future. If the physics of particles and fields is time symmetric and is properly defined only in the realm of the entire multiverse (or just in one of its finite-time universes if all the multiversal components are identical), then one must have a deep physical reason that makes the microscopic behaviour to appear as time symmetric. Such a physical reason can be found if we consider the existence of wormholes in the realm of the multiverse.

Since these wormholes accrete dark energy [95] they can actually grow so big that the whole spacetime of any of the universes making the multiverse is engulfed by the wormhole immediately before (after) the universe reaches (leaves) the future (past) big rip singularity. Such a gigantic process has been denoted as big trip and has hitherto been considered to just predict unwanted catastrophic cataclysms in the future of a hypothetical universe filled with phantom energy [95]. Nevertheless, when considered in the context of our multiversal model, the phenomenon of the big trip may provide unexpected benefits. If fact, it can be shown (see the preceding discussion in Chapter 8) that an observer from one of the finite-time universes that together compose the infinite multiverse would see big trips to crop up in his (her) future and his (her) past. Thus, such as it was pointed out in Ref. [136], the mouths of the wormholes in the past and future may be moving, and so travelling in time, in such a way that they can be inserted into each other right before the big trip so that the individual universe can freely journey from future to past and vice verse, so destroying any arrow of time of that universe and establishing a complete time symmetry in the whole multiverse. In Chapter 8 we have also discussed that a big trip phenomenon is prevented from taking place in a single universe with an infinite time, which therefore constitutes a reason for why there is an arrow of time in the observable universe.

The cropping up of wormholes with distinct sizes in the neighborhood of the big rips [123] of the multiverse helps us to furthermore consider a future event horizon for any observer in the multiverse, defined to have a proper size given by

$$R_h = a(t) \int_t^\infty \frac{dt'}{a(t')}. \tag{9.33}$$

Inserting then the expression for the scale factor given by Eq. (9.2) one can easily check that R_h becomes infinite for any observer in the multiverse, so making mathematically and physically fully consistent the consideration of any fundamental theory based on the definition of an S-matrix in the context of the multiverse. Since a single universe with infinite time resembles a de Sitter space and hence contains a *future event horizon* with finite proper size, it is not possible to define such fundamental theories in what we now consider a classical cosmology.

Based on the above considerations we can conjecture that the whole physical reality consists of a multiverse whose structure can for instance be described by the model reviewed in Sec. 9.2, which is a natural framework to consistently describe all time-symmetric particle physics and fields, with a universe we call our own being one of the infinitely many independent, identical universes that form up the multiverse. Such a singled component is perceived by observers in it as a space which currently expands in a super-accelerated fashion along an infinite cosmic irreversible time in a manner similar to a de Sitter space. If that conjectured description is adopted, then the known incompatibilities between particle physics and cosmology fade out. The price to pay would be a cosmological model with a negative energy density (which is nevertheless over-compensated by a positive cosmological constant in accordance with the requirements of quantum theory) which is defined on a physical domain where the quantization rules should be expected to be not the same as those used in the whole multiverse (or just in one of its finite-time universes if all the multiversal components are identical) for particles and fields.

9.4   The ruminations on the multiverse

We have tried to solve the incompatibilities between particle physics and cosmology by resorting to the *Copernican principle* that every founded scientific advance must necessarily be accompanied by taking the humankind off its lofty pedestal of uniqueness. The universe in which we reside appears to be merely one such from an infinite batch of other universe. If all (or at least the majority) of them share the observable properties of our own universe in the infinite cosmic time scale, then the resulting cosmic model will actually provide a number of tentative solutions to several key problems that arise when one tries to make compatible particle physics with current cosmology.

Starting with a multiversal model recently suggested, we have singled out a universe whose scale factor has been derived in terms of an infinite cosmic time. That universe is interpreted as being our own and is characterized by a total positive vacuum energy density which is made of two parts, a dynamical one which is small and negative and a positive cosmological constant. We have then been able to establish precise mathematical relations between the cosmological parameters of the multiverse and those of the observable universe which actually restore compatibility between cosmology and particle physics if the latter is taken to be defined in the original multiverse (or just in one of its finite-time universes if all the multiversal components are identical). Thus, we have shown that: (i) the ratio of values of the cosmological constant for particle physics and cosmology derived in this way is precisely what has been considered as a basis for formulating the so-called problem of the cosmological constant, (ii) because for an observer in any of the finite-time universes of the multiverse there are two big trip phenomena, one in the past and other in the future, all the physics in the multiverse will be time symmetric, but as no such phenomena may take place in the single infinite-time universe, there will be an arrow of time in our observable universe, and (iii) the proper size of the event horizon of the multiverse is infinite, i.e., there is no future event horizon in the multiverse, and therefore any fundamental physical description can be consistently carried out. All the above results amount to our main conjecture: If we assume that particle physics lives in the multiverse (or just in one of its finite-time universes if all the multiversal components are identical) where it is well defined, then our cosmology results from the observations performed in a framework of an infinite cosmic time of one of the infinite number of universes that form up the multiverse, with the known incompatibilities between particle physics and cosmology turning out to be nothing but artifacts arising from our attempt to interpret our own universe as containing everything.

Obviously a more realistic model should require the introduction of some matter. It is easy to see that if one adds a matter term to the Friedmann equation (9.1), this term would dominate only at small values of the scale factor, becoming negligible at the large values of a scale factor. Thus, our multiverse scenario, born out of infinitely many big rip singularities where the the size of the universes diverges, would still be valid.

We have seen how in many instances our considerations has required us to invoke the brane models. In the next chapter we shall finally take a closer look at these models and will see how the analytical methods we have developed in the course of this book helps to solve these models and in the process — to come up with some very interesting answers to many cosmological enigmas.

Part VI
Branes

Chapter 10

Nonsingular solutions on the brane and in the bulk

10.1 Introduction

In this chapter we will continue our quest for exact, analytic cosmological solutions. This time, however, we will turn our eye to the slightly more hypothetical — but much more exciting! — models of the *brane world*. These models consider our universe as a (3+1)-dimensional manifold embedded into a higher-dimensional *bulk space* — the latter typically being a 5-dimensional[1] manifold. Being originally introduced by *L. Randall and R. Sundrum* in [137], the model has since seen many revisals and enhancements. One of the more interesting of these newer models was presented in [138]; it describes the set of a parallel 3-branes, all of them being imbedded in five-dimensional bulk, filled with the gravitation and the scalar field. The action of set is expressed as:

$$S = \int d^4x dy \sqrt{|g|} \left(\frac{1}{2}R - \frac{(\nabla\phi)^2}{2} - V(\phi) \right)$$

$$- \sum_b \int_{y_b} d^4x \sqrt{|\tilde{g}^b|} \, |\sigma_b(\phi), \tag{10.1}$$

where (x^μ, y) with $0 \leq \mu \leq 3$ are the five-dimensional coordinates; R is the five-dimensional Ricci scalar; brane number b is located at $y = y_b$; g_{ab} is the 5-D metric, and $\tilde{g}^b_{\mu\nu}$ is the metric, induced on the b-th brane. (Note that, as usual, it is written in special system of units, where the gravitational constant $\kappa = 1$ and speed of light $c = 1$.) The tension on the bth brane is named σ_b, and the potential $V(\phi)$ is considered to be a function of the bulk scalar field $\phi = \phi(y)$. The field equations follows from the (10.1) by

---

[1]I.e. the (4+1)-dimensional.

the procedure of variation; their solution is assumed to have a form [138]:

$$ds^2 = dy^2 + A(y)\left(-dt^2 + e^{2Ht}\delta_{ik}dx^i dx^k\right), \tag{10.2}$$

where $A(y)$ is a sought warp factor, and $H = $ const is a Hubble parameter. If $H = 0$ *on the brane*, the brane will be the stationary one, and if $H > 0$, then the brane will be expanding in the de Sitter regime (cf. [139]–[140]).

The (10.1) leads to system of differential equations, that, in turn, can be rewritten in "supersymmetric" style by introduction of our old friend superpotential $W(\phi)$, defined here by the $\dot{\phi} \equiv W'(\phi)/2$ relation, where the dot denotes the derivative in y, and the prime — the derivative in ϕ. In case of $H = 0$ there exists following equation

$$V(\phi) = W'^2/8 - W^2/6. \tag{10.3}$$

Upon usage of this equation, the authors of [138] have suggested a simple way of obtaining of exact solutions of the field equations (see also [137], [141] for further details). Their approach can be expressed in following steps:

(i) Choosing convenient $V(\phi)$ and $W(\phi)$, making sure that the equation (10.3) with these new values still holds true;

(ii) Integration of equation $\dot{\phi} \equiv W'(\phi)/2$, in order to receive the relation $\phi = \phi(y)$;

(iii) When $H = 0$ there exists the relation $\dot{A}(y)/A(y) = -W(\phi(y))/3$. Hence, substitution of the obtained function for $\phi(y)$ in explicit equation for $W(\phi)$ and further integration results in the sought warp factor $A(y)$.

From this point, the famous solutions RS (Randall-Sundrum) turns out to be just a case $W(\phi) = \pm a^2 = $ const. Authors [138] have also considered the models with even $(W(\phi) = \pm \left(a^2 - b\phi^2\right))$, odd $(W(\phi) = \pm 2b^2\phi)$ and exponential $(W(\phi) = 2ae^{-k\phi})$ potentials (the even superpotentials had been previously discussed in [142], and the odd ones had been introduced in [143]).

The major advantages of the described method are it's simplicity and the fact that it allows for an easy construction of the nonsingular solutions (see also [143]). Lets consider the case of one brane, localized in $y = y_0$. From the equation $\dot{\phi} \equiv W'(\phi)/2$ it is possible to express y as a function of the field variable ϕ:

$$y(\phi) = y_0 + \int_{\phi_0}^{\phi} \frac{d\tilde{\phi}}{W'(\tilde{\phi})}, \tag{10.4}$$

where $\phi_0 = \phi(y_0)$ is a value of a scalar field on brane. It follows from the (10.4) that the very existence of zeros of the function $W'(\phi)$ ($\phi = \phi^\pm =$ const, $|\phi^\pm| < \infty$) is leading to the divergences of the integral. In another words, $\phi \to \phi^\pm$ when $y \to \pm\infty$. What it means, is that if we choose the superpotential as having at least two zeros at both $\phi > \phi_0$ and $\phi < \phi_0$ where W' is vanishing fast enough (for example $W'(\phi) \sim |\phi - \phi^\pm|^k$, with $k \geq 1$, and $\phi \to \phi^\pm$), then the field ϕ remains finite throughout all the *bulk* space. On the other hand, the case of $H = 0$ allows to express the Ricci scalar via the superpotential and its derivative in the form:

$$R(\phi) = 2(W'(\phi))^2/3 - 5W^2(\phi)/9, \tag{10.5}$$

thus, showing the finiteness of the scalar curvature for the aforementioned nonsingular superpotential for every y. The particular examples of the non-singular solutions for even, odd and exponential superpotentials can all be found in [138]. It also contains the generalization of this technique for the non-stationary branes ($H \neq 0$) and a brief discussion touching the possible usages of the method in solving such a well known problems as the problem of cosmological constant's smallness and the hierarchy problem.

However, in spite of all its aforementioned strong points, this method does have its weaknesses, the primal one being its "phenomenological" nature — the fact, that each physically consistent (i.e. leading to the reasonable $\phi(y)$ and $A(y)$) superpotential $W(\phi)$ can only be obtained by "guessing". Although all simple examples, considered in [138] were not too hard "to guess" and they do lead to sensible cosmological models, it would not necessarily be that easy, whenever we try the more difficult (and hence, more realistic) $W(\phi)$. These reasonings poses a principal question: does there exist such procedure of (10.1)'s nonsingular solutions construction, which is both mathematically simple and regular?..

To begin answering this question let us first take one further look at the relationships that exist between the superpotential $W(\phi)$, potential $V(\phi)$, the five-dimensional scale factor $A(y)$ and the parameter $\dot{\phi}$ for the special case $H = 0$ (see (10.3)):

$$\frac{d\ln A}{dy} = -\frac{1}{3}W,$$

$$\frac{d\phi}{dy} = \frac{1}{2}W'(\phi), \tag{10.6}$$

$$V(\phi) = -\frac{1}{6}W^2(\phi) + \frac{1}{8}(W'(\phi))^2.$$

This system looks remarkably familiar, doesn't it?.. This is because we have seen the system *exactly* like this one before, back in Sec. 6.1 when we have been discussing the classical Friedman-Robertson-Walker-Lemaître model! It was the system (6.8), that tied together the scalar field $x(t) = (3\sqrt{2}\alpha)^{-1}\varphi(t)$, its potential $V(x)$, the scale factor $a(t)$ (in terms of the Hubble parameter $H(t)$) and the *square root of the Friedman superpotential* $R(x)$. The system looked like this:

$$H = \frac{d\ln a}{dt} = \alpha R,$$

$$\frac{dx}{dt} = -2\sqrt{2}R'(x), \tag{10.7}$$

$$v(x) = (R(x))^2 - 4\left(R'(x)\right)^2,$$

and, as we recall from Sec. 6.2, it was this system that naturally lead us to the *Abel equation*. Of course, this must mean that the system (10.6) is also *reducible* to the Abel equation!.. However, there is one little observation we should make beforehand, and it deals with a little inconsistency in signs existing in the systems (10.6) and (10.7). It appears that in order to make the last equation of (10.6) identical to the last equation of (10.7), one should (among other things) necessarily switch from the function $V(\phi)$ to a new function $v(\phi)$ with the *opposite* sign. This requirement is not some mathematical trifle; its significance stems from the fundamental difference in the roles the superpotential W plays in the Friedman and the brane scenarios. Recall that in the Friedman model the superpotential has the form $W_f = 1/2\dot\varphi^2 + V(\varphi)$ and it has the physical meaning of *density* of the matter (energy) filling the Universe. But it is no longer the case for the superpotential for the brane model, as (10.6) clearly states that:

$$\frac{1}{6}W^2 = \frac{1}{2}\dot\phi^2 - V(\phi), \tag{10.8}$$

which means that the superpotential W here has the meaning of a *square root* of the **pressure** of the aforementioned matter. In the non-phantom scenario (i.e. when $\dot\phi^2 \neq 0$) the superpotential will always be real-valued, provided $V(\phi) < 0$. However, the fact that W is a *square root* of the pressure implies that the positive values of V are only allowed insofar as they are bonded from above by the inequality:

$$V(\phi) \leq \frac{1}{2}\dot\phi^2. \tag{10.9}$$

This inequality should not be violated, otherwise one will end up with an imaginary W and, hence, a non-real valued scale factor $A(y)$, which is, of course, unphysical. On the other hand, the phantomization is possible even in this model, but, again, only when $V < 0$ and is small enough to offset effect of the negative kinetic term $1/2\dot\phi^2$. In a way, the boundaries and corresponding scenarios appears to be the exact opposite of those from the Friedman scenario, hence a requirement to change a sign of the potential V in (10.6).

Knowing all this, we can now proceed with our calculations. Since the derivations will be completely analogous to those done in Sec. 6.2, we will omit them and jump right to the following theorem:

Theorem 10.1. *Let* $H = 0$, $x = 4/\sqrt{3}\,\phi$, $\chi = \ln|V|$, $\kappa = \pm 1$. *For a universe in a non-phantom zone for a given* $V(\phi) = v(x)$ *the corresponding superpotential* $W = W(x, C)$ *is defined as*

$$W(x, C) = y(x)\sqrt{\frac{6v(x)}{1 - y^2(x)}}, \tag{10.10}$$

where $y = y(x, C) \neq \pm 1$ *is a general solution of the Abel's equation of 1st kind:*

$$y' = -\frac{1}{2}\left(y^2 - 1\right)(\kappa - \chi'y). \tag{10.11}$$

The negative values of v *correspond to* $|y| > 1$ *and positive values of* v *correspond to* $|y| < 1$. *The case* $V = 0$ *occurs if and only if* $y = \pm 1$ *and the superpotential* W *has the form:*

$$W = Ce^{\kappa x/2}.$$

Furthermore, if the potential $V < 0$ *satisfies the condition* $\tilde v(\xi) = v(i\xi)$: $\mathbb{R} \to \mathbb{R}^-$ *for at least some* ξ, *for these values of* ξ *the universe can enter the phantom zone with superpotential* W:

$$\tilde W(x, C) = y(x)\sqrt{\frac{-6\tilde v(x)}{y^2(x) + 1}}, \tag{10.12}$$

where $y : \mathbb{R} \to \mathbb{R}$ *is the solution of the Abel equation*

$$\frac{dy}{d\xi} = \frac{1}{2}(y^2 + 1)(\kappa - \tilde\chi'y). \tag{10.13}$$

The Theorem 10.1 opens a whole slew of possibilities, as it not only provides us with an easy mechanism for a reconstruction of the previously known results (for example, the aforementioned Randall-Sundrum model can be produced by setting $V(\phi) = const$ and integrating the subsequent autonomous Abel equation) and for the generation of the new ones (including the phantom models!). Furthermore, since the core Eqs. (10.11) and (10.13) are identical to the Eqs. (6.24) and (6.26) that we have discussed in great detail in Chapter 6, we will not elaborate on them further and will simply refer the reader to the corresponding sections of the book.

Sadly, the method we have just described has but one serious flaw: it is only applicable for the special case of $H = 0$. In case of a non-zero Hubble parameter, the seminal system (10.6) is no longer true and has to be replaced with something more complicated. So, for the general case we will have to invent a different approach to the task at hand. For that we will turn to a very special way to approach the supersymmetry: the *Darboux Transformation*. The idea can be summarized in the following way. First of all, we known that the supersymmetric link between $V(\phi)$ and $W(\phi)$ is a typical one for the *supersymmetric quantum mechanics* (SSQM). In turn, SSQM is realizable through the usage of *Darboux transformation* (DT) for the *Schrödinger equation* [144–146], and that allows us to construct the integrable nonsingular potentials — providing that all the prop functions of *Crum's determinants* do have the intermittent zeros [144], [147]. So, it seems to be only natural trying to write the DT for the field equations directly, and then using them as a source of a new exact solutions. An existence of quite simple algebraic connections between the Ricci scalar and the super-potential (10.5) notes that construction of nonsingular potentials in SSQM via the right DT application will also aid in development of the systematical procedure of construction of the nonsingular solutions ($R \neq \infty$) in the brane theory, which, by that means, will turn out to be a regular one — as distinct from the "phenomenological approach" of [138]. In particular, the DT method, if applied to just a single exactly solvable potential, grants us the infinite number of an exact nonsingular solutions, with the mentioned potential being a "fuse" in all those solutions construction. For example, in this work we'll take the RS for the role of such a fuse solution. Another interesting way of the DT usage is a construction of the chains of discrete symmetries and the further inquiries of their closures. Such a technique allows getting the new exact solutions, which are manifested through the Painlevé transcendents and the higher extensions.

Let us now see how these ideas work together. We begin by discussing the virtual crux of our method: the *Darboux transformation* and how to define them on the brane, beginning with the simplest case of $H = 0$.

10.2 From the Schrödinger equation to the jump conditions

We begin with a case $H = 0$. We already know from the Theorem 10.1 that in this case a radical simplification to the Able equation is possible. However, for the sake of the argument we will abscond from using it and will instead try to reduce our cosmological equations to a *linear* differential equation. For this end we introduce the new variable $u \equiv \dot{A}/A$, then use the (10.1) and (10.2) to get the following set of equations [138]:

$$\dot{u} = -\frac{2\dot{\phi}^2}{3} - \frac{2}{3}\sum_b \sigma_b(\phi)\delta(y - y_b), \quad u^2 = \frac{\dot{\phi}^2}{3} - \frac{2}{3}V(\phi). \tag{10.14}$$

The third equation has the form

$$\ddot{\phi} + 2u\dot{\phi} = \frac{\partial V(\phi)}{\partial \phi} + \sum_b \frac{\partial \sigma_b(\phi)}{\partial \phi}\delta(y - y_b),$$

but it is not independent and follows from the system (10.14).

Suppose we a single brane, located at $y = 0$. Then $\sigma(\phi(y))\delta(y) = \sigma_0\delta(y)$, where $\sigma_0 = \sigma(\phi(0))$ (supposing $\phi(0) \neq 0$). It is now very easy to check that the function $A^2(y)$ satisfies the *Schrödinger equation*

$$\left(\frac{d^2}{dy^2} + \frac{8}{3}V(\phi(y)) + \beta\delta(y)\right)A^2(y) = 0, \tag{10.15}$$

where $\beta = 4\sigma_0/3$. We are therefore ready to define the spectral problem

$$\ddot{\psi} = (v(y) - \beta\delta(y) - \lambda)\,\psi, \tag{10.16}$$

where $\psi = \psi(y; \lambda)$ and $v(y)$ is a certain potential which will be further referred to as a quantum-mechanical potential, or QM-potential. Let's assume that it is possible to solve the Eq. (10.16) formally (i.e., obtaining among others some "non-physical" solutions which do not belong to L^2) for some QM-potential $v(y)$ and all $\lambda \in \mathbb{R}$. Suppose also that for some values of the spectral parameter λ (say, $\lambda = \lambda_0$) the solutions of the (10.16) will be strictly positive: $\psi(y; \lambda_0) > 0$. In this case we can conclude that function $A(y; \lambda_0) = \sqrt{\psi(y; \lambda_0)}$ is the solution of the (10.15) with the potential

$$V(\phi(y)) = \frac{3}{8}(\lambda_0 - v(y)). \tag{10.17}$$

Therefore, the Eq. (10.16) can, in fact, be considered a generator of the exact solutions of (10.15). Substituting $A(y)$ and $V(\phi(y))$ (derived from the (10.16)) into (10.14) we'll find out $\phi = \phi(y)$. For specificity, we'll further assume the following condition for the potential $v(y)$:

$$v(y) = \text{const} + \tilde{v}(y), \qquad \int_{-\infty}^{+\infty} dy \, (1 + |y|) \, |\tilde{v}(y)| < \infty. \qquad (10.18)$$

It is also worth mentioning that the condition (10.18) is not essential for our calculations, and is used here merely for the sake of convenience.

Suppose now that we have managed to determine the parameters $\psi(y)$, $v(y)$ and λ. As a next step we should find the functions $A(y)$, $\phi(y)$ and $V(\phi)$. In order to do this we'll examine the region located far from the brane to exterminate the δ-function term in the (10.16) from further considerations. Quite obviously, the quantity $A(y)$ can only be restored if $\psi(y)$ is nonnegative for all y. The unique wave function that satisfies this condition and also belongs to the physical spectrum of the Schrödinger equation (10.16), is called a wave function of ground state with $\lambda = \lambda_{vac}$. We should also mention that there exists another possible way to choose the "good" wave function, namely, to consider the case $\lambda = \lambda_{-1} < \lambda_{vac}$. This satisfies the imposed condition, since the (10.16) indeed admits the always positive solutions $\psi(y; \lambda_{-1})$. These solutions obviously do not belong to the L_2 space and definitely do not describe the bound states, but the warp factor $A = \sqrt{\psi(y; \lambda_1)}$ will nevertheless be real. However, these solutions result in singularity at infinity: $A \to \infty$ at $y \to \infty$, and therefore the ground state $\psi(y; \lambda_{vac})$ is by far the best alternative to $\psi(y; \lambda_{-1})$

It is a well known fact that DT has a remarkable capacity to enable the engineering of ad hoc potentials with arbitrary finite discrete spectrum, thus it can be used as a regular procedure for the construction of the system (10.14)'s exact solutions. We'll now quickly remind the very essence of the *DT method* [148]. Let $\psi_1 = \psi(y; \lambda_1)$ and $\psi_2 = \psi(y; \lambda_2)$ be the two solutions of the Eq. (10.16), and let also $\psi_1 > 0$ for every y it is defined on. The ψ_1 is called the prop function of Darboux transformation. In general, the transformation law is:

$$\psi_2 \to \psi_2^{(1)} = \frac{\dot{\psi}_2 \psi_1 - \dot{\psi}_1 \psi_2}{\psi_1},$$

$$v \to v^{(1)} = v - 2 \frac{d^2}{dy^2} \log \psi_1, \qquad \lambda_2 \to \lambda_2. \qquad (10.19)$$

Darboux transformation (10.19) is an isospectral symmetry of Eq. (10.16), because the dressed function $\psi_2^{(1)}$ is the solution of the dressed

Eq. (10.16):

$$\ddot{\psi}_2^{(1)} = \left(v^{(1)} - \beta^{(1)}\delta(y) - \lambda_2\right)\psi_2^{(1)},$$

with a new (dressed) potential $v^{(1)}$ but with the same value of $\lambda_2^{(1)} = \lambda_2$. The transformation law for the prop function is given by the relation ([144]):

$$\psi_1 \to \psi_1^{(1)} = \frac{1}{\psi_1}\left(C_1 + C_2 \int \psi_1^2(y)dy\right),$$

$$\ddot{\psi}_1^{(1)} = \left(v^{(1)} - \beta^{(1)}\delta(y) - \lambda_1\right)\psi_1^{(1)},$$

(10.20)

with arbitrary constants $C_{1,2}$ (and, of course, $\lambda_1^{(1)} = \lambda_1$). The most interesting part for us here is a ground state of a dressed Hamiltonian. There exist three possibilities.

1) If ψ_1 is a wave function of a ground state, and λ_2 is an energy of a first excited level, then the new dressed function $\psi_2^{(1)}$ will be the wave function of a new Hamiltonian's ground state [145], [146]. All other values of the discrete spectrum of this Hamiltonian are received by simple obliteration of level λ_1 from the initial spectrum.

2) The prop function $\psi_1 = \psi_1(y; \lambda_1)$ does not belong to an L_2, but it does not have any zeros either, and $\psi_1 \to e^{+p^2|y|}$ when $y \to \pm\infty$. In the case, after the DT we'll get the new Hamiltonian, whose discrete spectrum is just an old one except for the new state $\lambda = \lambda_1$ added, thereby being the new ground state. The corresponding wave function $\psi_1^{(1)}$ (10.20) with $C_2 = 0$ becomes the ground state's wave function [149].

3) The aforementioned prop function can in fact be represented as a linear superposition of two positive, linearly independent solutions $\psi_1^{(\pm)}$ with the following characteristics: $\psi_1^{(\pm)} \to 0$ when $y \to \mp\infty$ and $\psi_1^{(\pm)} \to e^{+p^2|y|}$ when $y \to \pm\infty$. Taking either $\psi_1^{(+)}$ or $\psi_1^{(-)}$ as a prop function results in creation of a new Hamiltonian, whose discrete spectrum totally coincides the initial one [149].

Needless to say, we are interested primarily in the second case.

As a next step of our consideration we should examine the jump conditions imposed by the brane. These conditions can be written as

$$\psi_{i+}(0) = \psi_{i-}(0) = \psi_i(0), \qquad \dot{\psi}_{i+}(0) - \dot{\psi}_{i-}(0) = -\beta\psi_i(0). \qquad (10.21)$$

where $\psi_{i+}(y)$ and $\psi_{i-}(y)$ are the solutions of Schrödinger equation (10.16), both with the same potential and eigenvalues, but valid only for positive or negative y correspondingly:

$$\ddot{\psi}_{i\pm} = (v(y) - \lambda_i)\,\psi_{i\pm}, \qquad (10.22)$$

It is easily verified, that (10.21) are exactly the same jump conditions as in before-cited article [138]:

$$u_+(0) - u_-(0) = \frac{1}{2}\left(\frac{\dot{\psi}_+(0)}{\psi_+(0)} - \frac{\dot{\psi}_-(0)}{\psi_-(0)}\right)$$

$$= \frac{\dot{\psi}_+(0) - \dot{\psi}_-(0)}{2\psi(0)} = -\frac{\beta}{2} = -\frac{2\sigma_0}{3},$$

being, in essence, the Israel conditions [150], [151]. The next theorem proves, that DT keeps the jump conditions (10.21) unchanged:

Theorem 10.2. *If $\psi_i(y) = \{\psi_{i+}(y > 0), \psi_{i-}(y < 0)\}$ with $i = 1,2$ are solutions of (10.22), satisfying the (10.21) and if $\psi_1 = \{\psi_{1+}, \psi_{1-}\}$ is a prop function, then the transformed function $\psi_2^{(1)} = \{\psi_{2+}^{(1)}, \psi_{2-}^{(1)}\}$ (where the prop functions of $\psi_{2\pm}$ are $\psi_{1\pm}$ correspondingly) satisfies the same conditions (10.21) with $\beta^{(1)} = -\beta$:*

$$\psi_{2+}^{(1)}(0) = \psi_{2-}^{(1)}(0) \equiv \psi_2^{(1)}(0),$$
$$\dot{\psi}_{2+}^{(1)}(0) - \dot{\psi}_{2-}^{(1)}(0) = +\beta\psi_i(0). \qquad (10.23)$$

The proof of the theorem is conducted by the direct calculation. We'll just cite two formulas, which turned to be quite useful in proving:

$$\dot{\tau}_{1\pm} = v - \lambda_1 - \tau_{1\pm}^2, \qquad \dot{\psi}_{2\pm}^{(1)} = (\lambda_1 - \lambda_2)\psi_{2\pm} - \tau_{1\pm}\psi_{2\pm}^{(1)},$$

where $\tau_{1\pm} = \dot{\psi}_{1\pm}/\psi_{1\pm}$. An obvious result of this theorem is the conclusion, that n-times repeated DT leads to the dressed functions, satisfying the (10.21) with tension $\beta^{(n)} = (-1)^n\beta$, on the assumption that n prop functions-solutions $\{\psi_{i\pm}\}$ ($i = 1,..,n$) of the initial Schrödinger equation (10.22) and one transformed solution $\psi_{n+1\pm} = \psi$ with the same potential, but different λ are all satisfy the condition (10.21). Such n-times dressing is usually convenient to introduce via the *Crum's formulas* [152]:

$$A_\pm^{(n)}(y) = \sqrt{\frac{D_n^{(\pm)}(y)}{\Delta_n^{(\pm)}(y)}}, \qquad v_\pm^{(n)}(y) = v - 2\frac{d^2}{dy^2}\log\Delta_n^{(\pm)}(y), \qquad (10.24)$$

where

$$
\Delta_n^{(\pm)}(y) = \begin{vmatrix} \psi_{n\pm}^{[n-1]} & \psi_{n\pm}^{[n-2]} & \cdots & \psi_{n\pm} \\ \psi_{n-1\pm}^{[n-1]} & \psi_{n-1\pm}^{[n-2]} & \cdots & \psi_{n-1\pm} \\ & \star & & \\ & \star & & \\ \psi_{2\pm}^{[n-1]} & \psi_{2\pm}^{[n-2]} & \cdots & \psi_{2\pm} \\ \psi_{1\pm}^{[n-1]} & \psi_{1\pm}^{[n-2]} & \cdots & \psi_{1\pm} \end{vmatrix},
$$

$$
D_n^{(\pm)}(y) = \begin{vmatrix} \psi_{\pm}^{[n]} & \psi_{\pm}^{[n-1]} & \psi_{\pm}^{[n-2]} & \cdots & \psi_{\pm} \\ \psi_{n\pm}^{[n]} & \psi_{n\pm}^{[n-1]} & \psi_{n\pm}^{[n-2]} & \cdots & \psi_{n\pm} \\ \psi_{n-1\pm}^{[n]} & \psi_{n-1\pm}^{[n-1]} & \psi_{n-1\pm}^{[n-2]} & \cdots & \psi_{n-1\pm} \\ & \star & & & \\ & \star & & & \\ \psi_{2\pm}^{[n]} & \psi_{2\pm}^{[n-1]} & \psi_{2\pm}^{[n-2]} & \cdots & \psi_{2\pm} \\ \psi_{1\pm}^{[n]} & \psi_{1\pm}^{[n-1]} & \psi_{1\pm}^{[n-2]} & \cdots & \psi_{1\pm} \end{vmatrix},
$$

and $\psi^{[k]} \equiv d^k \psi / dy^k$. These formulas are very useful in our study. In particular, the solution, written in this form, taken together with (10.23), allows us to extend the conditions, introduced by [138]:

$$
u_+^{(n)}(0) - u_-^{(n)}(0) = (-1)^{n+1} \frac{2\sigma_0}{3},
$$

where $u_\pm^{(n)} = \dot{A}_\pm^{(n)}/A_\pm^{(n)}$ and σ_0 is a tension on the initial brane. Moreover, using the (10.24) in the procedure of levels addition one can show, that the wave function of ground state for potential $v_\pm^{(n)}$ is defined by the formula:

$$
\psi_{n\pm}^{(n)} = \left(A_\pm^{(n)} \right)^2 = \frac{\Delta_{n-1}^{(\pm)}}{\Delta_n^{(\pm)}}. \tag{10.25}
$$

Wave function of a ground state is not equal to zero, and this statement is similar to condition $A_\pm^{(n)} \neq 0$. Using the n-times dressed scale factor, we can calculate the corresponding Ricci scalar:

$$
R_\pm^{(n)} = -\frac{4\ddot{A}_\pm^{(n)} A_\pm^{(n)} + \left(\dot{A}_\pm^{(n)} \right)^2}{(A_\pm^{(n)})^2}, \tag{10.26}
$$

where the $\pm$ signs are referring to $y > 0$ (over the brane) and $y < 0$ (under the brane). Since the denominator of this expression is always non-zero, it follows right from the (10.25) and (10.26) that Ricci scalar remains finite not just on brane, but also in a bulk. If, for example, the behavior of $\psi^{(n)}_{n\pm}$ is an asymptotical one, i.e. if $\psi^{(n)} \to e^{-p^2|y|}$ when $y \to \pm\infty$, then $R^{(n)}_{\pm} \to -5p^4$ at infinity. Thus, in the case of single stationary brane, the DT method allows the simple construction of an abundant set of an exact nonsingular solutions of Eq. (10.14).

Remark 10.1. As we have seen, the Eq. (10.16) can be extremely effective in producing of the exact solutions of (10.14). In the cases of exactly solvable potentials $v(y)$, the usage of DT allows the construction of a large (in fact, infinite) set of such solutions. The problem arises if the consideration of the Schrödinger equation begins with an arbitrary solution of the (10.14). This solution gives us, in turn, the single exact solution of the (10.15) with $v(y) = -8V(\phi(y))/3$ and $\lambda = 0$ — but, generally, it is not guaranteed, that $v(y) = -8V(\phi(y))/3$ will be one of exactly solvable potentials, and hence — not provided that we can solve the spectral problem (10.16) for all other admitted λ. This means, that, in order to use the DT method, we should in first place carefully choose the initial solution of the (10.14). It is highly remarkable, that the famous RS model actually does the trick!

10.3   Dressing the Randall-Sundrum brane solution

To illustrate the effectiveness of formulas (10.19), (10.20) we'll take the example of the RS brane dressing, which case corresponds to a simple quantum potential $v = \mu^2 = \text{const}$. Here $\tilde{v} = 0$ (see (10.18)), and the potential of a scalar field $V = -3\mu^2/8$. Solution ψ_1 of Eq. (10.16) with eigenvalue $\lambda_1 = 0$ has the form

$$\psi_{1\pm} = a_{1\pm}e^{\mu y} + b_{1\pm}e^{-\mu y}, \quad \psi_{1+} = \psi_1(y > 0),$$
$$\psi_{1-} = \psi_1(y < 0), \tag{10.27}$$

where $a_{1\pm}$, and $b_{1\pm}$ are real positive constants. Usage of (10.21) results in following equations:

$$a_{1-} = a_{1+} + \frac{\beta(a_{1+} + b_{1+})}{2\mu}, \quad b_{1-} = b_{1+} - \frac{\beta(a_{1+} + b_{1+})}{2\mu}. \tag{10.28}$$

It is obvious from (10.28) that $a_{1+} + b_{1+} = a_{1-} + b_{1-}$, therefore it is convenient to take advantage of the parameterization:

$$a_{1\pm} = r^2 \sin^2 \alpha_\pm, \quad b_{1\pm} = r^2 \cos^2 \alpha_\pm. \tag{10.29}$$

(10.29) and (10.28) allows us to get the compatibility conditions:

$$- \sin^2 \alpha_+ < \frac{\beta}{2\mu} < \cos^2 \alpha_+, \qquad (10.30)$$

where $|\beta| < 2\mu$ necessarily. Thus, in order to get the new exact solutions via DT, starting out from the RS solutions, we should use (10.27) combined with addition conditions (10.28) and (10.30).

Now, let's take (10.27) as a prop function. After the DT's (10.19) execution we get

$$\mu^2 - \beta\delta(y) \to \mu^2 + v_\pm^{(1)} - \beta^{(1)}\delta(y),$$

where

$$v_\pm^{(1)} = -\frac{8\mu^2 a_{1\pm} b_{1\pm}}{\left(a_{1\pm} e^{\mu y} + b_{1\pm} e^{-\mu y}\right)^2},$$

while $v_+^{(1)} = v^{(1)}(y > 0)$ and $v_-^{(1)} = v^{(1)}(y < 0)$. This potential suffers a jump on the brane:

$$v_+^{(1)}(0) - v_-^{(1)}(0) = -4\mu\beta \left(\frac{\beta}{2\mu} + \frac{a_{1+} - b_{1+}}{a_{1+} + b_{1+}} \right),$$

which is equally zero if the tension is

$$\beta = \frac{2\mu(b_{1+} - a_{1+})}{a_{1+} + b_{1+}}. \qquad (10.31)$$

This case is really special for it allows to get the potential $v_-^{(1)}$ from $v_+^{(1)}$ by simple permutation of a_{1+} and b_{1+}. Let's choose $\psi_{1\pm}^{(1)} = 1/\psi_{1\pm}$.[2] It is clear, that for $y \to 0$:

$$\dot\psi_{1+}^{(1)}(0) - \dot\psi_{1-}^{(1)}(0) = -\beta^{(1)}\psi_1^{(1)}(0) = +\beta\psi_1^{(1)}(0),$$

which, according to (10.23) is exactly the way it had to be. Using the $\psi_{1\pm}^{(1)}$ we receive a new metric

$$A_\pm^{(1)} = \frac{1}{\sqrt{a_{1\pm} e^{\mu y} + b_{1\pm} e^{-\mu y}}}.$$

It's asymptotical behavior is: $A^{(1)}(y) \to e^{-\mu y/2}/\sqrt{a_{1+}}$, when $y \to +\infty$, and $A^{(1)}(y) \to e^{\mu y/2}/\sqrt{b_{1-}}$, when $y \to -\infty$. (10.31) is a particular case of

---

[2]Thereby, defining the integration's constants $C_1 = 1$ and $C_2 = 0$, which is distinct for a ground state.

$a_{1+} = b_{1-}$, thus

$$A^{(1)}(y) \rightarrow \frac{1}{\sqrt{a_{1+}}} e^{-\mu|y|/2}, \qquad y \rightarrow \pm\infty.$$

If we additionally insert condition $b_{1+} = a_{1-} = 0$, we will get, after the substitution $\beta \rightarrow -\beta$, the reduction of $A^{(1)}$ to the solution of the RS set (10.14), with the potential being equivalent to the initial $V = -3\mu^2/8$.

Returning to the general case, we receive

$$\left(\dot{\phi}_{\pm}^{(1)}\right)^2 = \frac{3\mu^2 a_{1\pm} b_{1\pm}}{\left(a_{1\pm} e^{\mu y} + b_{1\pm} e^{-\mu y}\right)^2},$$

and, after the simple calculation, we come to well-known sine-Gordon potential:

$$V^{(1)}(\phi_{\pm}) = -\frac{3\mu^2}{8} \cos\left[\frac{4}{\sqrt{3}}(\phi_{\pm}(y) - \phi_{0\pm})\right],$$

where

$$\phi(y)_{\pm} = \phi_{0\pm} + \sqrt{3} \arctan\left(\sqrt{\frac{a_{1\pm}}{b_{1\pm}}} e^{\mu y}\right),$$

and $\phi_{0\pm}$ are the arbitrary constants.

Finally, to make the picture complete, we'll derive the quantity $\left(\sigma_0^{(1)}\right)'$:

$$\left(\sigma_0^{(1)}\right)' = \dot{\phi}_+^{(1)}(0) - \dot{\phi}_-^{(1)}(0)$$

$$= \mu\sqrt{3} \sin\left(\alpha_+ - \alpha_-\right) \cos\left(\alpha_+ + \alpha_-\right).$$

Note, that in the case (10.31) we have

$$\beta = 2\mu \cos 2\alpha_+, \qquad \alpha_- = \alpha_+ + \frac{\pi}{2}(2N+1),$$

where N is a whole number.

10.4   The generalization of the method

This section is dedicated to three questions, each of them being connected with the further development and generalization of the method; namely: the $n = 2$ dressing of the RS solutions, the role of the shape-invariance and the two interesting ways of generalization of the model with the odd superpotential. The first way is realizable via the *Adler theorem* whereas for the second one the dressing chains of discrete symmetries should be used.

Let's return to the RS model. The solution of it's Schrödinger equation with eigenvalue $\lambda_2 = \mu^2 - \nu^2$ can be chosen in form:

$$\psi_{2\pm} = a_{2\pm} e^{\nu y} + b_{2\pm} e^{-\nu y}, \quad \psi_{2+} = \psi_1(y > 0),$$
$$\psi_{2-} = \psi_1(y < 0). \tag{10.32}$$

Let's assume

$$a_{2-} = a_{2+} + \frac{\beta(a_{2+} + b_{2+})}{2\nu}, \quad b_{2-} = b_{2+} - \frac{\beta(a_{2+} + b_{2+})}{2\nu}. \tag{10.33}$$

Using (10.19) for (10.32) we get

$$\psi_{2\pm}^{(1)} = \frac{\Omega_{\pm}}{a_{1\pm} e^{\mu y} + b_{1\pm} e^{-\mu y}},$$
$$\Omega_{\pm} = (\nu - \mu)\left(a_{1\pm} a_{2\pm} e^{(\mu+\nu)y} - b_{1\pm} b_{2\pm} e^{-(\mu+\nu)y}\right) \tag{10.34}$$
$$+ (\nu + \mu)\left(a_{2\pm} b_{1\pm} e^{(\nu-\mu)y} - a_{1\pm} b_{2\pm} e^{(\mu-\nu)y}\right).$$

Function (10.34) will be used as a prop function for second DT. For the construction of a new ground state with energy $\lambda_2 < 0$ we should choose $\nu > \mu > 0$, $a_{2\pm} > 0$ and $b_{2\pm} < 0$ — this way the function $\psi_{2\pm}^{(1)}$ will not have the unwanted zeros. Consideration of (10.33) gives us the conclusion, that value of positive coefficient $a_{2\pm}$ should satisfy the inequalities:

$$\frac{a_{2+}}{a_{2-}} > \left(\frac{2\nu}{2\nu + \beta}\right)_{\beta>0}, \quad \frac{a_{2-}}{a_{2+}} > \left(\frac{2\nu}{2\nu - \beta}\right)_{\beta<0}.$$

If these relations are correct, we are free to use (10.24) and (10.25) for the second DT, which results in:

$$v_{\pm}^{(2)} = \mu^2 - 2\frac{d^2}{dy^2} \log \Omega_{\pm}, \quad A_{\pm}^{(2)} = \sqrt{\frac{a_{1\pm} e^{\mu y} + b_{1\pm} e^{-\mu y}}{\Omega_{\pm}}}.$$

It is quite easy to make sure, that what we get is a brane with correct jump condition and the tension $\sigma_0^{(2)} = +\sigma_0$. Also note, that potential $v^{(1)}$ has two levels: $\lambda_1 = 0$ and $\lambda_2 = \mu^2 - \nu^2$.

It appears, that the proper work of DT method is only provided for exactly solvable potentials. One of the efficient ways of obtaining of such potentials lies in a usage of so-called shape invariants [153]. If an initial potential is a function of y and some free parameters a_i: $v = v(y; a_i)$, and after the DT one gets $v^{(1)} = v^{(1)}(y; a_i^{(1)})$ then v is called the *shape-invariant (SI) potential*. SI-potentials are quite common in quantum mechanics, e.g. the harmonic oscillator [144]. The major point here is that exactly solvable potentials from [138] (for the models without cosmological expansion) are also the SI-potentials. We'll confine ourselves with considering the three examples from the cited article, retaining the terminology.

A. Exponential potential. Superpotential is $W(\phi) = 2be^{-\phi}$, where $\phi(y) = \log(a - by)$. Therefore

$$v(y) = \frac{c}{(by - a)^2},$$

which is a well known SI potential: $v^{(1)} = \text{const} \times v$.

B. Even superpotential. For the model [142] we get (with $A(0) = 1$):

$$\log A(y) = -\frac{ay}{3} + g\left(1 - e^{-2by}\right).$$

In this case

$$v(y) = \frac{8}{3}gb(2a - 3b)e^{-2by} + 16g^2b^2e^{-4by} + \frac{4a^2}{9}.$$

After the DT

$$v \to v^{(1)} = v - 4\frac{d^2}{dy^2}\log A,$$

we get

$$v^{(1)}(y) = \frac{8}{3}gb(2a + 3b)e^{-2by} + 16g^2b^2e^{-4by} + \frac{4a^2}{9}.$$

Thus $v^{(1)}$ can be obtained from v by substitution $a \to -a$ and $g \to -g$. It means that $v(y)$ is an SP-potential.

C. Odd superpotential. In this case we have

$$\log A(y) = -ay - by^2.$$

A calculation yields

$$v(y) = 4\left(2by + a\right)^2 - 4b.$$

It is nothing else, but the harmonic oscillator and, hence, an SP-potential: $v^{(1)} = v + \text{const}$. Let's consider this example as the fuse in further constructing of the new exact solutions via DT.

10.5 The Darboux transformation and the harmonic oscillator

As we have seen, the odd superpotential $W(\phi(y)) = 2\kappa b^2 \phi(y)$, with $\kappa = +1$ ($y > 0$) or $\kappa = -1$ ($y < 0$) results in

$$\phi(y) = \phi_0 + \kappa b^2 y,$$

$$\log A(y) = \log A_0 + \frac{\phi_0^2 - \phi^2(y)}{3}, \tag{10.35}$$

where $A_0 = A(0)$, $\phi_0 = \phi(0)$. Using (10.35) we get the Schrödinger equation of the harmonic oscillator

$$v(y; \kappa) = \frac{16}{9} b^4 \left(\kappa b^2 y + \phi_0\right)^2.$$

Its ground state has the wave function $\psi_1(y; \kappa) = e^{-2(\kappa b^2 y + \phi_0)^2/3}$, and energy $\lambda_1 = 4b^4/3$. Since $\psi_1(0; +1) = \psi_1(0; -1)$ and

$$\dot{\psi}_1(0; +1) - \dot{\psi}_1(0; -1) = -\beta e^{-2\phi_0^2/3},$$

the tension will be $\beta = 8b^2 \phi_0/3$. All other eigenfunctions and eigenvalues can be obtained via the formulas

$$\psi_n(y; \kappa) = \left(\partial_y - 4b^2(b^2 y + \kappa\phi_0)/3\right)^{n-1} \psi_1(y; \kappa),$$

$$\lambda_n = 4b^2(2n - 1)/3. \tag{10.36}$$

In the article [147] Adler has suggested the more general way of deletion of a groups of an excited states. It ensures the general state λ_{vac} of a transformed Hamiltonian to stay the same, though being described by a totally new wave function. Using this theorem and (10.24) one can delete the even number of adjacent levels (if the number will be odd, for example $N = 1$ like before, then the new potential appears to be singular). Upon deletion of levels λ_3 and λ_4 one gets *Higgs-like potential* $v^{(2)}(y; \kappa)$, with the ground state $\psi_1^{(2)}(y; \kappa)$ and the tension $\beta^{(2)}$:

$$v^{(2)} = \frac{16b^4}{9} \frac{4096x^{10} + 12288x^8 + 21888x^6 + 10368x^4 - 22599x^2 + 2187}{(64x^4 + 27)^2},$$

$$\psi_1^{(2)} = \frac{8x^2 + 3}{64x^4 + 27} e^{-2x^2/3}$$

$$\beta^{(2)} = \frac{8b^2 \phi_0(512\phi_0^6 + 960\phi_0^4 + 792\phi_0^2 - 243)}{3(64\phi_0^4 + 27)(8\phi_0^2 + 3)},$$

$$\tag{10.37}$$

where $x = \kappa b^2 y + \phi_0$. The potential $v^{(2)}$ is represented on the Fig. 10.1.

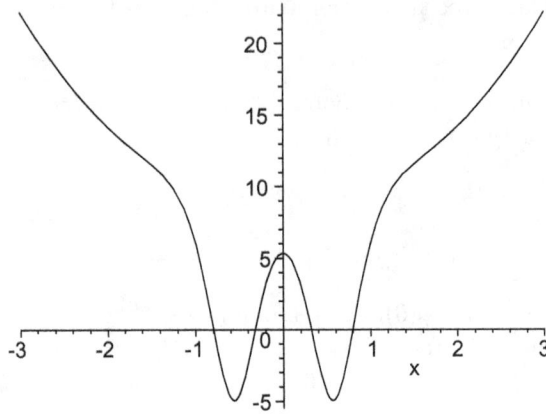

Fig. 10.1   The harmonic oscillator without levels λ_2 and λ_3.

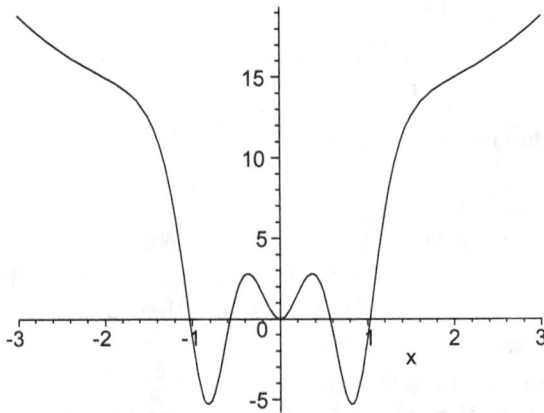

Fig. 10.2   The harmonic oscillator without levels λ_3, λ_4, λ_6, λ_7.

After the deletion of levels λ_3, λ_4, λ_6 and λ_7 we receive one very interesting potential which is shown on Fig. 10.2. Finally, the Fig. 10.3 represents the potential which is obtained through the elimination of the levels $\lambda_4 - \lambda_7$. The expressions for the potentials $v^{(n)}(y; \kappa)$, eigenfunctions (including the ground state) and tensions $\beta^{(n)}$ can be easily obtained via the (10.24) and (10.36). We omit them only for their extreme bulkiness.

The reason why these potentials have such Higgs-like form is evident. In fact, the spectrum of potential with two symmetric minimums necessarily

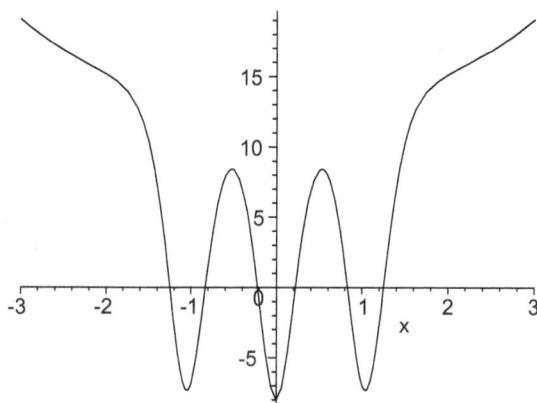

Fig. 10.3   The harmonic oscillator without levels $\lambda_4 - \lambda_7$.

contains the lacuna between two down levels and all other spectrum. Deleting the levels λ_3 and λ_4 we construct the spectrum with such lacuna so it is only natural that this procedure results in nothing else but Higgs-like potential. In a similar manner, elimination of the levels $\lambda_4 - \lambda_7$ leads to the spectrum with lacuna between the three bottom levels and the other spectrum. As a result one get the potential with three minimums (Fig. 10.3). Note that all these potentials have the following asymptotic: $v^{(n)}(y; \kappa) \to 16b^8 y^2/9$, when $y \to \pm\infty$.

Now lets return to potential (10.37). Using the exact forms of the $v^{(2)}$ and $\psi_1^{(2)}$ one can obtain the potential $V(\phi)$ in parametric form (i.e. $V = V(y)$ and $\phi = \phi(y)$). The plot of the function $d\phi(x)/dx$ for $y > 0$ is represented on the Fig. 10.4. This expression is real for $\mid x \mid \geq x_* \sim 0.25735$ so if we identify the $\phi_0 = x_*$ then the field $\phi(y)$ will be well determined all around the bulk.

10.6   The dressing chains and the fourth Painleve equation as the generalization of an odd superpotential

Another interesting generalization of the odd superpotential (i.e. the harmonic oscillator) can be obtained with the aid of the dressing chains of the DT, which chains were previously introduced in [154]. Let's suppose, that we have defined the set of functions $f_k = \partial_y \log \psi_k^{(k-1)}$, where $\psi_k^{(n)}$ are n-times dressed solutions of the (10.16). Then these functions will also be

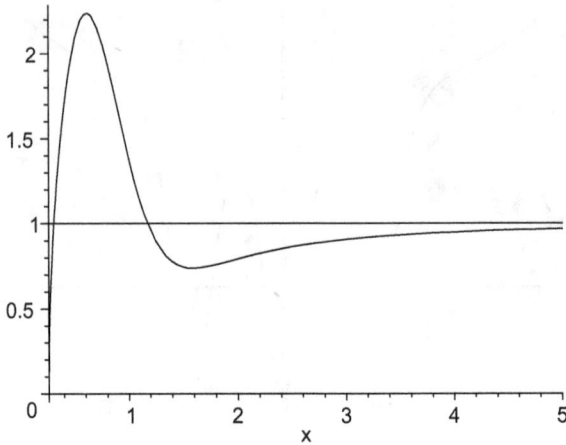

Fig. 10.4   The function $\phi'(x)$.

solutions of the *dressing chain*:

$$\dot{f}_k + \dot{f}_{k+1} = f_k^2 - f_{k+1}^2 + \alpha_k. \qquad (10.38)$$

where α_k are constants, and they can be expressed via spectral parameters λ_k. Following [154] we'll consider the periodic version $f_k = f_{k+N}$, $\alpha_k = \alpha_{k+N}$ with positive integer N which will be further referred as period. The whole theories of dressing chains are totally different for the odd ($N = 2n + 1$) and even ($N = 2n$) periods. The same is valid for the cases with $\alpha = 0$ and $\alpha \neq 0$, where

$$\alpha = \sum_{k=1}^{N} \alpha_k.$$

If $\alpha \neq 0$ and $N = 1$ we get the harmonic oscillator. In the case $N = 3$ ($\alpha \neq 0$) we get the fourth *Painlevé equation* (P_{IV}) for the function $P(y) = -f_1 - y$:

$$\ddot{P} = \frac{\dot{P}^2}{2P} + \frac{3}{2}P^3 + 4yP^2 + 2\left(y^2 - a\right)P + \frac{b}{P}, \qquad (10.39)$$

where $\alpha = 2$, $a = -\alpha_1 - \alpha_2/2 + 1$, $b = -\alpha_2^2/2$. Thus the potential $v(y) = \dot{f}_1 + f_1^2$ can be expressed in term of the *Painlevé-IV transcendents* and has the following behavior

$$v(y - y_0) \sim \frac{\alpha^2 y^2}{36} + My \cos\left(\frac{\alpha y^2}{\sqrt{3}} + \delta_0\right). \qquad (10.40)$$

Using (10.40) one can conclude that this potential is indeed the generalization of the harmonic oscillator. In particular, this potential is growing quadratically.

We should also mention, that the spectrum of the $v(y)$ can be found explicitly [154]. Moreover, it can be shown that the ground state is at $\lambda_1 = 0$.

10.7 The models with the cosmological expansion

If we are going to study the models with cosmological expansion, we should expose the DT method to some kind of generalization. This requirement inevitably follows from the fact, that the substitution $A^2 = \psi$ now results in nonlinear equation rather then in linear Schrödinger equation. However, seemingly being as difficult as it seems, this equation still can be reformulated in the way of linear spectral problem which does admit the Darboux transformation.

10.7.1 *Noncompact spacetimes*

If in (10.2) $H \neq 0$ then instead of (10.14) we have a much more complex system:

$$\dot{u} = -\frac{2\dot{\phi}^2}{3} - \frac{2H^2}{A} - \frac{2}{3}\sum_{b=1}^{N}\sigma_b\delta(y - y_b),$$

$$u^2 = \frac{\dot{\phi}^2}{3} - \frac{2}{3}V(\phi) + \frac{4H^2}{A},$$

$$(10.41)$$

written for the set of N branes which are located at $y = y_b$.

The function A^2 solves the following nonlinear equation

$$\left(-\frac{d^2}{dy^2} - \frac{8}{3}V(\phi(y)) + \frac{12H^2}{A} - \frac{4}{3}\sum_{b=1}^{N}\sigma_b\delta(y - y_b)\right)A^2 = 0. \qquad (10.42)$$

Therefore, we can formally introduce the spectral problem (the Schrödinger equation):

$$\ddot{\psi}(y;\lambda) = \left(U(y) - \sum_{b=1}^{N}\beta_b\delta(y - y_b) - \lambda\right)\psi(y;\lambda), \qquad (10.43)$$

where

$$U(y) = -\frac{8}{3}V(\phi(y);\lambda) + \frac{12H^2}{\sqrt{\psi(y;\lambda)}}. \qquad (10.44)$$

We stress here, that $U(y)$ is not a function of the spectral parameter λ whereas V and ψ are. With this induced, we are free to use the spectral theory for the equation (10.43). For any given $U(y)$ one can solve (10.43) to get ψ and to find V from (10.44). Of course, at the end of our solution we should, in order to find $A = \sqrt{\psi}$, receive ψ of positive value, but this problem is also avoidable, as has been demonstrated in Secs. 10.3 and 10.4. Indeed, one can construct positive solution via the DT starting out from any simple solvable model — taking the RS model for example. There is, however, one interesting difference with the case of stationary brane: the "kinetic term" appears to have the form

$$\frac{\dot{\phi}^2}{3} = \frac{1}{4}\left(\left(\frac{\dot{\psi}}{\psi}\right)^2 - U + \lambda\right) - \frac{H^2}{\sqrt{\psi}}, \qquad (10.45)$$

therefore we can't just use the ground state solution like we did in the case of stationary brane, or at $y \to \infty$ wave function $\psi \to 0$ and we will get $\dot{\phi}^2 < 0$. Obviously, this problem requires a more through examination. So, let's again start out from an "RS potential" $U = \mu^2$, and consider the case $\lambda = 0$. The solution will have the form (10.27). Using it as the prop function in (10.19), one gets the same metric $A_{\pm}^{(1)}$ from Sec. 10.3, only with another potential. The main problem is now focused in the kinetic energy of the scalar field, having the form

$$(\dot{\phi}_{\pm}^{(1)})^2 = \frac{3\mu^2 a_{\pm} b_{\pm}}{\left(a_{\pm}e^{\mu y} + b_{\pm}e^{-\mu y}\right)^2} - 3H^2\sqrt{a_{\pm}e^{\mu y} + b_{\pm}e^{-\mu y}}.$$

It is clear that for large enough y, one gets $\dot{\phi}^2 < 0$. On the other hand, if (see (10.29))

$$\left(\frac{H}{\mu}\right)^2 < \frac{\sin^2 2\alpha_{\pm}}{4r},$$

then on the brane and in its vicinity, there are no such troubles. Therefore the simplest way of avoiding the problem is just dealing with the finite volume, which is less then the minimum value of y, resulting in the negativity of the kinetic term.

If we nevertheless wish to deal with infinite volume, we had to be sure to use only those positive solutions $\psi(y) = \{\psi_+(y)$, at $y > 0$; $\psi_-(y)$, at $y < 0\}$ which are growing as $|y| \to \infty$. The construction of such solutions can be done in different manners. For example, one can take the solution of the *Schrödinger equation* with $\lambda = \lambda_{-1} < \lambda_{vac}$. The more interesting opportunity lies in applying of so called B-potential (Bargmann potentials)

$U(y)$ whose solutions of the *Schrödinger equation* all have the form: $\psi = F(y)\exp(\alpha y)$ where $F(y)$ is some polynomial [155]. To show them in work lets rewrite the equation (10.45) in the form

$$\frac{1}{3}\dot{\phi}_{\pm}^2 = -\frac{1}{2}\dot{u}_{\pm} - \frac{H^2}{\sqrt{\psi_{\pm}}}, \tag{10.46}$$

where $u_{\pm} = \dot{\psi}_{\pm}/(2\psi_{\pm})$. The necessary criterion here is that $\dot{u}_{\pm} < 0$ for any given y. Let's choose the $\psi_{\pm}$ in typical "Bargmann form":

$$\psi_{\pm} = \left(Ay^2 \pm By + C\right)e^{\pm\alpha y}, \tag{10.47}$$

where α, A, B and C are positive constants, associated by the condition

$$2AC < B^2 < 4AC. \tag{10.48}$$

The jump condition (10.21), imposed to single brane at $y = 0$ results in $\beta = -2(B + \alpha C)/C$. The solution of the equation (10.41) appears to be

$$A_{\pm} = e^{\pm\alpha y/2}\sqrt{Ay^2 \pm By + C}. \tag{10.49}$$

where $A(y) = A_+$ at $y > 0$ and $A(y) = A_-$ at $y < 0$. The condition (10.48) guarantees that this solution will be nonsingular on brane as well as in bulk (since $A(y) \neq 0$ for any values of y) and also that $\dot{u}_+ < 0$ for $y > 0$ and $\dot{u}_- < 0$ for $y < 0$. Finally, choosing

$$H^2 < \frac{B^2 - 2AC}{4C\sqrt{C}},$$

and using (10.49) and (10.46) one get

$$\frac{\dot{\phi}_{\pm}^2}{3} = \frac{2A^2y^2 \pm 2ABy + B^2 - 2AC}{4(Ay^2 \pm By + C)^2} - \frac{H^2 e^{\mp\alpha y/2}}{\sqrt{Ay^2 \pm By + C}},$$

where $\phi(y) = \phi_+$ at $y > 0$ and $\phi(y) = \phi_-$ at $y < 0$. Now we can see, that it is always possible to find large enough values of α that will greatly decrease the second member of equation with regard to the first, thus leaving the whole equation positive for both $y > 0$ and $y < 0$.

10.7.2 Orbifold model

In previous sections, our efforts were wholly concentrated on spacetimes that are noncompact in y. Now we'll consider the compactified case. As we shall see, the DT in this case allows to obtain the solutions, resulting in exponentially small 4-D effective cosmological constant.

Let's choose the potential in (10.43) as $U = \mu^2 \sin^2 \theta$. We'll start with the two solution of this equation:

$$\psi_1 = \cosh(\mu \sin \theta y), \quad \psi_{2\pm} = \sinh(\mu y \pm b),$$

where $b = \text{const}$. The function ψ_1 is the solution of the (10.43) with $\lambda_1 = 0$ whereas $\psi_{2\pm}$ with $\lambda_2 = -\mu^2 \cos^2 \theta$. Using ψ_1 as the prop function we can dress $\psi_{2\pm}$ with the help of the DT (10.19). The result will be

$$\psi_{2\pm}^{(1)} = \cosh(\mu y \pm b) - \sin \theta \tanh(\mu \sin \theta y) \sinh(\mu y \pm b). \tag{10.50}$$

where we should choose the sign "$+$" for $y > 0$ and sign "$-$" for $y < 0$. One can show that $\psi_{2\pm}^{(1)} > 0$ for all y. In this case the dressed potential $U^{(1)}$ will have the single bound state with $\lambda = \lambda_1$ and eigenfunction $1/\psi_1$.

Now we can compactify the y direction as an $\mathbb{Z}_2$ *orbifold*, with two branes sitting at each of the two fixed points ($y = 0$ and $y = L$). The new aspect here is that $S_1/\mathbb{Z}_2$ orbifold model permits only one bulk space. Due to the symmetry of the model, we only have to consider the jump conditions at $y = 0$ and $y = L$. The direct calculations result in a very simple expression for the tension $\sigma_0 = -3\mu \tanh b \cos^2 \theta / 2$ for the brane which is located at $y = 0$ (we'll choose it as the visible brane), while the tension on $y = L$ brane ends up looking considerably more complex:

$$\sigma_L = -\frac{3\mu}{2F_L(\theta)} \Big(\sinh(\mu L + b) \big(\cosh^2(\mu \sin \theta L) - \sin^2 \theta\big)$$
$$- \sin \theta \sinh(\mu \sin \theta L) \cosh(\mu \sin \theta L) \cosh(\mu L + b) \Big),$$
$$F_L(\theta) = \cosh(\mu \sin \theta L) \Big(\sin \theta \sinh(\mu \sin \theta L) \sinh(\mu L + b) \tag{10.51}$$
$$- \cosh(\mu \sin \theta L) \cosh(\mu L + b) \Big)$$

Both of these expressions can, however, be greatly simplified by the choice $b = -\mu L$:

$$\sigma_0 = \frac{3\mu}{2} \tanh(\mu L) \cos^2 \theta,$$

$$\sigma_L = -\frac{3\mu}{2} \tanh(\mu \sin \theta L) \sin \theta.$$

Since the expansion rate on the visible brane is $H(0) = H/\sqrt{A(0)} = H/\left(\psi_{2\pm}^{(1)}(0)\right)^{1/4}$ then

$$H(0) = \frac{H}{\sqrt[4]{\cosh(\mu L)}}.$$

Therefore, for large values of μL the effective 4-D cosmological constant Λ_{eff} on the visible brane is

$$\Lambda_{eff} \sim H^2(0) = \frac{H^2}{\sqrt{\cosh(\mu L)}} \sim H^2 e^{-\mu L/2}. \tag{10.52}$$

This means, that 4-dimensional cosmological constant, as seen by observers from the visible brane, becomes exponentially small while the L grows large, therefore, automatically solving the cosmological constant problem right in the framework of the the brane worlds models (this idea was first suggested in article of S.-H.H. Tye and I. Wasserman [141]).

All we need now is to make sure, that $\dot\phi_+^2 > 0$ at $0 < y < L$ (since we have the single bulk between two branes then it is enough to consider the case $0 < y < L$ solely). Let's choose $\theta \sim \pi/2$. In this case $\lambda_2 \sim \lambda_1 = 0$. We stress, however, that $\theta \neq \pi/2$ exactly so $\lambda_2 = -\mu^2\alpha^2$ where $\alpha = \pi/2 - \theta \ll 1$. To obtain $\dot\phi_+^2(0) > 0$ it is necessary that $\cosh(\mu L) > 16H^4/\mu^4$ whereas $\dot\phi_+^2(L) > 0$ if $\cosh(\mu L) < \mu/(2H)$. The combination of these conditions is:

$$\left(\frac{2H}{\mu}\right)^4 < \cosh(\mu L) < \frac{\mu}{2H},$$

which is possible if $\mu > 2H$. In fact, it is better to imply an enhanced condition $\mu \gg 2H$, because large values of μ allows to avoid the conclusion that $\sigma_0 = 3\mu\alpha^2/2 \sim 0$. Since μ is not bounded from above, we draw the wanted conclusion, that our model indeed permits positive $\dot\phi_+^2$ at $0 < y < L$.

Nota Bene 3. Of course, the described theory could not generally protect some finite number of regions (at $0 < y < L$) from special situations when $\dot\phi_+^2(y) < 0$ (if the value of μ is not sufficiently large, for example). In this case, however, one can use the method which was supposed in [92], viz. these regions can be cut out, with their boundaries being sewn together in such a way that neither the warp factor (along with its first two derivatives) nor Ricci scalar will experience a jump. This unexpected fortune raises from the fact that matching is done at an inflection point of $\log A$ (more correctly, this is approximately true if ψ is sufficiently large to neglect the term $H/\sqrt{\psi}$ in either one of equations (10.45) and (10.46)).

10.8 The Green function method

This is another way to construct solution for the models with cosmological expansion. First of all, we can directly transform the system (10.41) into the linear equation rather than use substitution $\psi = A^2$. Indeed, introducing the potential

$$U \equiv -\frac{2}{3}\left(\frac{\dot{\phi}^2}{2} + V(\phi)\right),$$

we get (for the case of single brane located at $t = 0$)

$$-\ddot{A} + \left(U - \frac{2}{3}\sigma_0\delta(y)\right)A = -2H^2. \tag{10.53}$$

Solving linear nonhomogeneous equation (10.53) for given U we can find $\phi(y)$ and $V(\phi)$ via the system

$$\dot{\phi}^2 = -\frac{3}{2}\left(U - \left(\frac{\dot{A}}{A}\right)^2\right) - \frac{6H^2}{A},$$

$$V = -\frac{3}{4}\left(U + \left(\frac{\dot{A}}{A}\right)^2\right) + \frac{3H^2}{A}.$$

Now, instead of the (10.53) let's consider the nonhomogeneous spectral problem

$$-\ddot{\psi}(y;\lambda) + \left(U(y) - \lambda - \frac{2}{3}\sigma_0\delta(y)\right)\psi(y;\lambda) = -2H^2.$$

which can be written as

$$\mathbf{L}(y;\lambda)\psi(y;\lambda) = -2H^2, \tag{10.54}$$

with $\mathbf{L}(y;\lambda) = d^2/dy^2 + U(y) - \lambda - 2\sigma_0\delta(y)/3$.

To solve (10.54) we introduce the homogeneous linear equation

$$\mathbf{L}(y;\lambda)\psi_0(y;\lambda) = 0, \tag{10.55}$$

and the *Green function* $G(y, z; \lambda)$:

$$\mathbf{L}(y;\lambda)G(y, z; \lambda) = \delta(y - z).$$

By solving these equation for given $U(y)$ and λ we can construct the solution of (10.54) as

$$\psi(y;\lambda) = \psi_0(y;\lambda) - 2H^2\int dz\, G(y, z; \lambda).$$

If $\psi(y;\lambda)$ will be positive then we will be able to define $A = \sqrt{\psi}$ just like before. But first we have to determine those conditions which make such positive solutions possible. This is still the open question. We should only note, that Green function can be constructed via solutions of the (10.55).

If initial potential $U(y)$ is exactly solvable one, then the same is true for N-times Darboux dressed potential $U^{(N)}$. Thus, if we know the Green function $G(y, z; \lambda)$ of the initial equation then we know the Green function $G^{(N)}(y, z; \lambda)$ for equation with the dressed potential. In other words, the DT is appliable here, at least formally.

10.9 A few notes on the possible generalizations

This approach can be easily generalized to the case of K-branes. For this we should replace the term $\beta\delta(y)$ in (10.16) by

$$\sum_b \beta_b \delta(y - y_b),$$

and also replace the two jump conditions (10.21) by $2K$ jump conditions at $y = y_b$. This is nothing more than technical matter, that's why we omit the details.

Another one kettle of fish is that even in case of general position, where we cannot reconstruct the form of the potential $V(\phi)$ as a function of ϕ, all these solutions will have the RS asymptotics (of course, if constructed precisely according to Sec. 10.3). The DT gives a link between the solvable problems, and probably most (if not all) of the exactly solvable potentials, from the potential of the harmonic oscillator to the finite-gap potentials [154], can all be obtained via these transformations. The physical sense of these potentials in the brane world is not clear. But our main purpose was to merely demonstrate and advertise the DT as being the very powerful tool for manufacturing the exactly solvable potentials in the 5-D gravity with the bulk scalar field. It has been demonstrated that this method can be useful both in models with and without the cosmological expansion, although in the first case the situation is more complex and deserves the further studying. Notably, there is one imperfection of this method: the DT would work for the 5-D gravity only. If $D > 5$ then the warp factor is a multi-variable function. Unfortunately, there is no cogent and effective general DT theory in several dimensions [156]. But, nevertheless, for $D = 5$ the DT is by far seems to be the best way to construct the nonsingular exact solutions.

Chapter 11

The singular brane solutions with a finite scale factor

11.1 Singularities on the brane: the classification

Starting out from the discovery of the cosmic acceleration [157] there have been constructed many models of the dark energy, including the very unusual ones: the phantom energy, the tachyon cosmologies, the brane worlds etc. Consideration of these models results in some unexpected conclusions about possibility of new cosmological doomsday scenarios: the *Big Rip singularity* (BRS) [158], the *Big Freeze singularity* (BFS) [159], [160], the *Sudden Future singularity* (SFS) [161], the *Big Boost singularity* (BBtS) [162], and the *Big Break singularity* (BBS) [163], [164]. In all these models the evolution ends with the curvature *singularity*, $|\ddot{a}(t)| \to \infty$, reachable in a finite proper time, say as $t \to t_s$. BRS and BFS both take place in the phantom models but with the different equations of state. In particular, BRS takes place if $w = p/\rho = \text{const} < -1$ whereas BFS occurs for the dark energy in the form of a phantom generalized *Chaplygin gas*. Models with the SFS, BFS, BBtS and BBS singularities are characterized by a finite value of the cosmological radius but different values of Hubble expansion parameter $H_s = H(t_s)$ and different signs of (divergent) expression $\ddot{a}_s/a$ (cf. also [165]):

$$a_s = \infty, \quad H_s = +\infty, \quad \frac{\ddot{a}_s}{a_s} = +\infty, \qquad \text{(BRS)}$$

$$a_s < \infty, \quad H_s = +\infty, \quad \frac{\ddot{a}_s}{a_s} = +\infty, \qquad \text{(BFS)}$$

$$a_s < \infty, \quad 0 < H_s < \infty, \quad \frac{\ddot{a}_s}{a_s} = -\infty, \qquad \text{(SFS)}$$

$$a_s < \infty, \quad 0 < H_s < \infty, \quad \frac{\ddot{a}_s}{a_s} = +\infty, \qquad \text{(BBtS)}$$

$$a_s < \infty, \quad H_s = 0, \quad \frac{\ddot{a}_s}{a_s} = -\infty. \qquad \text{(BBS)}$$

Remark 11.1. One of *classifications of singularities* for the modified gravity was given in [166]; for the classification and discussion concerned with avoiding the singularities in the alternative gravity dark energy models cf. [167]. Another classification of finite-time future singularities (Type I-IV singularities) is presented in [168]. According to this classification, the BRS is a singularity of type I, BFS is of type III, SFS and BBtS are type II and BBS — type II with $\rho_s \equiv \rho(t_s) = 0$ (although this is a quite non-trivial special case of a type II singularities). Our classification doesn't contain singularities of IV type (for $t \to t_s$, $a \to a_s$, $\rho \to 0$, $|p| \to 0$ and higher derivatives of H diverge) but as we shall see in section 11.6, the classification of Ref. [168] is not exactly complete too: the type IV is the special case of a more general type of singularities.

Remark 11.2. Another type of "singularity" — so called *w-singularity* was obtained in [169]. This "singularity" has a finite scale factor, vanishing energy density and pressure, and the singular behavior manifesting itself only in a time-dependent barotropic index $w(t)$. The w-singularities seem to be most similar to the type IV but are different nonetheless since they do not lead to any divergence of higher order derivatives of H [169].

One surmises that *w-singularity* is not a correct physical singularity since all the physical values (i.e. density, pressure and higher derivatives of the scale factor or Hubble roots) are finite. Moreover, the definition of *w-singularity* from the [169] is an incomplete one. To show this let us consider the following form of the scale factor

$$a(t) = a_s - A \left(t_s - t \right)^m. \qquad (11.1)$$

(11.1) is the special case of the general form of the scale factor from the [169] (with $B = 0$, $A = a_s$, $C/t_s^n = -A$, $D = 1$, $n = m$). One can show that for $t \to t_s$:

(a) type III singularity if $0 < m < 1$;
(b) type II singularity if $1 < m < 2$;
(c) w-singularity if $m > 2$.

The case $m = 1$ results in a model with the constant barotropic index $w = -1/3$. The case $m = 2$ is the most interesting one because

$$\rho \to 0, \quad p \to \frac{4A}{3a_s} \neq 0, \quad |w| \to \infty.$$

and

$$\frac{d^{2n}H}{dt^{2n}} = 0, \quad \frac{d^{2n+1}H}{dt^{2n+1}} \sim \frac{A^{n+1}}{a_s^{n+1}} < \infty,$$

at $t = t_s$. Thus we have some generalization of w-*singularity* such that the pressure is non-vanishing and finite at $t = t_s$.

The BRS and BFS have been obtained in the phantom cosmologies (BRS for the phantom perfect fluid with equation of state $p/\rho = w = \text{const} < -1$ and BFS for the phantom Chaplygin models. Throughout this chapter we'll stick to the metric units with $8\pi G/3 = c = 1$). The BBtS is connected to the effect of the conformal anomaly that drives the expansion of the Universe to a maximal value of the Hubble constant, after which the solution becomes complex. The BBS takes place in tachyon models.

Unlike BRS, BFS and BBtS altogether, the BBS and SFS are violating just the dominant energy condition ($\rho \geq 0$, $-\rho \leq p \leq \rho$). It is also possible to obtain some generalization of these singularities. In particular, the generalization of the Sudden Future singularities (the so called Generalized Sudden Future singularities or GSFS) would be have the derivative of pressure $p^{(m-2)}$ diverge, while being accompanies by the blow-up of the m-th derivative of the scale factor $a^{(m)}$ [170]. These singularities are possible in theories with higher-order curvature quantum corrections [168] and correspond to the classification introduced in this chapter.

Despite the fact that there has recently been a great inflow of articles, elaborating on the aforementioned singularities, an absolute majority of them has been of a mathematical nature, while the physical reasons for arousal of such singularities still remain less then clear. A remarkable exception is the article [160], which has introduced for the first time a new type of cosmological singularities located on the brane (for discussion about the soft singularities on brane with the quantum corrections cf. [171]). These singularities are characterized by the fact that while the Hubble parameter and scale factor remain finite, all higher derivatives of the scale factor ($\ddot{a}$ etc.) diverge as the cosmological singularity is approached. These singularities may be obtained as the result of embedding of $(3+1)$-dimensional brane in the bulk and this is why these singularities will be henceforth referred to as the "brane-like" singularities. We'll define the "brane-like" singularities in a following fashion: we'll say that *singularity is of a "brane-like" type if at the instance of its occurrence both scale factor and density remain finite and nonzero, while all the higher order derivatives of scale factor (starting with the second order) become altogether singular, i.e. $a \to a_s$, $\rho \to \rho_s$, $0 < a_s < \infty$, $0 < \rho_s < \infty$, $d^n a_s / dt^n = \infty$ for $n > 1$.*

Evidently, the class of "brane-like" singularities includes the singularities of Type II (with $\rho_s \neq 0$) or SFS and BBtS. Moreover, BBS will also be of this type whenever we are talking about the models with the constant positive curvature, since at the singularity point $\rho_s = 1/a_s^2$.

The physical nature of "brane-like" singularities emergence is quite clear: in the simple case with Z_2 reflection symmetry and the identical cosmological constants on the two sides of the brane, the dynamical equation contains few additional terms. One of them is the square root of the sum of contributions of density (on the brane), tension, cosmological constant and the "dark radiation" (the last one arises due to the projection of the bulk gravitational degrees of freedom onto the brane [160]). This sum is not positively defined and might become negative during the cosmological evolution. Thus, the solution of the cosmological equations can't be continued beyond the point where this sum turns to zero and what we end up with at this point is nothing but a "brane-like" singularity. Since the existence of such singularities is natural in the brane physics, it won't be against the logic to assume that the appearance of "brane-like" singularities in usual Friedmann cosmology (SFS or BBtS) might be an evidence of validity of the brane hypothesis. Therefore it is interesting to consider "brane-like" singularities without a brane (i.e. in FLRW cosmology) to establish the particular form of potential and the equation of state that will result in such singularities during cosmological dynamics. Such potential and the equation of state may altogether be useful for answering the big cosmological question: Don't we really live on the brane?

Furthermore, such singularities may actually result in very unusual models. In fact, let's consider the universe which contains a "brane-like" singularity. If the universe is filled with a scalar field while the Hubble parameter $H(t_s) = H_s$ and the scale factor $a(t_s) = a_s$ are finite at the singular point ($H_s < \infty$, $a_s < \infty$) then the value of the scalar field $\phi(t_s) = \phi_s$ might be finite as well. On the other hand, quantum corrections of higher order (in N-loops approximation) depend on the higher derivatives. If higher derivatives of scale factor diverge then this will also be the case for the scalar field. So one can expect that since all higher derivatives of scale factor and field alike diverge as the cosmological singularity is approached, then the quantum effects will be dominating for $t \to t_s$. This will be the case in spite of the fact that both density ρ_s and scale factor a_s will be finite and that ρ_s might be small and a_s — very large.

It may seem that quantum corrections will be dominating because the pressure $|p| \to \infty$ as the cosmological singularity is approached. This is not the case for the singularities of the IV type being "brane-like" by definition. Moreover, in Sec. 11.6 we'll construct the singularities of even more general type that will violate the classifications of [168] since ρ_s and p_s will be finite and nonzero and all higher derivatives will diverge.

We are going to construct the "brane-like" singularities in the FLRW universe filled by the usual self-acting, minimally coupled scalar field or homogeneous tachyon field. In this cosmology we'll also construct the singularities (with finite scale factor) where that Hubble variable vanishes and all higher derivatives of the scale factor diverge as the cosmological singularity is approached. That type of singularities is the generalization of the Big Break singularities and there will also be those of the "brane-like" type for the case of a constant positive curvature. We will calculate both self-acting potential $V(\phi)$ and tachyon potential $V(T)$ that result in appearance of such singularities. Additionally, we'll present the equation of state for such models. Moreover, the classification of singularities from the Ref. [168] will be complemented.

We'll also discuss a simple yet useful method for construction of exact solutions of the cosmological Friedmann equations with both self-acting, minimally coupled scalar field and a homogeneous tachyon field. The method itself will be denoted as the method of linearization and it has been previously suggested in [172] (see also [24], [74], [173]). We'll give a brief discussion of this method in the next Section.

11.2   The linearization method

Let us write the Friedmann equations as[1]

$$\left(\frac{\dot{a}}{a}\right)^2 = \rho - \frac{k}{a^2}, \quad 2\frac{\ddot{a}}{a} = -(\rho + 3p). \tag{11.2}$$

The crucial point here is the fact that a Friedmann equations admits a linearizing substitution and can therefore be studied via the different

---

[1]For the purpose of brevity in this section we will omit the dependence of the energy density ρ and pressure p on the scalar field φ.

powerful mathematical methods which were specifically developed for the linear differential equations. This is the reason we call our approach the method of linearization [172]:

Proposition. *Let $a = a(t)$ (with $p = p(t)$, $\rho = \rho(t)$) be a solution of (11.2). Then for the case $k = 0$ the function $\psi_n \equiv a^n$ is the solution of the Schrödinger equation*

$$\frac{\ddot{\psi}_n}{\psi_n} = U_n, \tag{11.3}$$

with potential

$$U_n = n^2 \rho - \frac{3n}{2}(\rho + p). \tag{11.4}$$

For example:

$$U_1 = -\frac{\rho + 3p}{2}, \quad U_2 = \rho - 3p, \quad U_3 = \frac{9}{2}(\rho - p),$$

or

$$U_{1/2} = -\frac{1}{2}\left(\rho + \frac{3p}{2}\right), \quad U_{-1} = \frac{5\rho + 3p}{2}$$

and so on.

Remark 11.3. If the universe is filled with scalar field ϕ whose Lagrangian is

$$L = \frac{\dot{\phi}^2}{2} - V(\phi) = K - V, \tag{11.5}$$

then the expression (11.4) will be

$$U_n = n(n-3)K + n^2 V.$$

In particular case $n = 3$ $U_3 = 9V(\phi)$. This particular case has been extensively studied in [24], [74].

Remark 11.4. For small values of $n \ll 1$ one gets $U_n \sim -3n(\rho+p)/2$; for example, if $n = 0.01$ then

$$U_n \sim -(0.0149\rho + 0.015p) \sim -0.015(\rho + p).$$

Therefore one can use $U_n < 0$ to check whether the weak energy condition is violated.[2] If, on the contrary, $n \gg 1$ then $U_n \sim n^2\rho$.

---

[2] At the same time, one shall keep in mind that this equation is nothing but approximate. To ensure (with the help of ψ_n) that the weak energy condition will indeed be violated, one can use exact equation $\dot{\sigma}_n = -3n(\rho + p)/2$, where $\sigma_n = \dot{\psi}_n/\psi_n$.

Remark 11.5. If $k = \pm 1$ then the Proposition is valid only for the case $n = 0, 1$.

As shown in [173], the representation of the Einstein-Friedmann equations as a second-order linear differential Eq. (11.3) allows for a usage of an arbitrary (known) solution for construction of another, more general solution parameterized by a set of $3N$ constants, where N is an arbitrary natural number. The large number of free parameters should prove itself useful for constructing a theoretical model that agrees satisfactorily with the results of astronomical observations. In particular, $N = 3$ solutions in the general case already exhibit inflationary regimes [173]. Unlike the previously studied two-parameter solutions (see [24], [74]), these three-parameter solutions might describe an exit from inflation without any fine tuning of parameters as well as the several consecutive inflationary regimes.

In the next section we will show that the method of linearization is indeed an effective one for construction of a "brane-like" singularity.

11.3 The "Brane-like" singularity

Assume

$$U(t) = \frac{\kappa u_s^2}{(t_s - t)^\alpha}, \tag{11.6}$$

with $u_s^2 = \text{const} > 0$, $\kappa = \pm 1$, $\alpha > 0$. For simplicity one has omitted the index n: $U_n \to U(t)$, $\psi_n \to \psi$. For $|t - t_s| \gg 1$ the potential $U(t) \to 0$ so $\psi(t) \sim t$ and $a(t) \sim t^{1/n}$. If $n = 2$ we have a universe filled with radiation ($w = 1/3$), and for $n = 3/2$ we have a dust universe with $w = 0$.

Now let us consider the solution of the (11.3) at $t \to t_s$:

$$\psi(t) = \psi_s + \sum_{k=1}^{\infty} c_k \left(t_s - t\right)^{k(2-\alpha)}, \tag{11.7}$$

where

$$\psi_s = a_s^{1/n} = \frac{(2-\alpha)(\alpha-1)}{\kappa u_s^2}, \tag{11.8}$$

and

$$c_k = \frac{\kappa^{k-1} u_s^{2(k-1)}}{(2-\alpha)^{k-1} k! \prod_{m=0}^{k-2} \left[(k-m)(2-\alpha) - 1\right]}, \tag{11.9}$$

with $c_1 = 1$. For $\alpha < 2$ the series (11.7) is convergent for any t (including $t = t_s$) since its radius is

$$R = \lim_{k \to \infty} \left| \frac{c_k}{c_{k+1}} \right| = +\infty.$$

If $\alpha > 2$ then the first two terms of (11.7) are

$$\psi = \psi_s + \frac{1}{(t_s - t)^{\alpha-2}},$$

so the function $\psi(t)$ will be singular at $t \to t_s$ and we have either Big Rip ($n > 0$) or Big Crunch ($n < 0$) at $t = t_s$. If $\alpha = 2$ the general solution of the (11.3) will be of the form

$$\psi = \sqrt{t_s - t}\left(C_1(t_s - t)^L + C_2(t_s - t)^{-L}\right), \tag{11.10}$$

with the arbitrary constants $C_{1,2}$ and $L = \sqrt{1 + 4\kappa u_s^2}/2$. For any C_1 and C_2 the solution (11.10) results in Big Rip or Big Crunch singularity as well. Finally, if $\alpha = 1$ then the series (11.7) will be

$$\psi = (t_s - t)\left[1 + \frac{\kappa u_s^2}{2}(t_s - t) + \frac{u_s^4}{12}(t_s - t)^2 \right.$$
$$\left. + \frac{\kappa u_s^6}{144}(t_s - t)^3 + \frac{u_s^8}{2880}(t_s - t)^4 + \cdots\right],$$

so there is no cosmological singularity and a is finite.

Thus, in framework of our investigation one must consider only $0 < \alpha < 1$ and $1 < \alpha < 2$. Keeping only the two first terms in (11.7) and using

$$H = \frac{\dot{\psi}}{n\psi}, \qquad \frac{\ddot{a}}{a} = \frac{\ddot{\psi}}{n\psi} - (n-1)H^2, \tag{11.11}$$

one gets

$$\psi \sim \frac{(2-\alpha)(\alpha-1)}{\kappa u_s^2} - (t_s - t)^{2-\alpha},$$

$$H = \frac{\kappa u_s^2}{n(\alpha-1)}(t_s - t)^{1-\alpha}, \tag{11.12}$$

$$\frac{\ddot{a}}{a} = \frac{u_s^2}{n(t_s - t)^\alpha}\left[\kappa - \frac{u_s^2(n-1)}{n(\alpha-1)^2}(t_s - t)^{2-\alpha}\right].$$

Using (11.12) one can show that the conditions $\psi_s > 0$ and $H > 0$ will hold if $n > 0$. So one has two cases:

(i) If $\kappa = -1$ then $0 < \alpha < 1$ and one gets BBS;
(ii) if $\kappa = +1$ then $1 < \alpha < 2$ and one gets BFS.

To obtain SFS one should use another solution of (11.3):

$$\psi = \psi_s - n\psi_s H_s(t_s - t) + \frac{\kappa u_s^2 \psi_s}{(2-\alpha)(1-\alpha)} (t_s - t)^{2-\alpha}$$

$$+ \sum_{k=1}^{\infty} c_{2k} (t_s - t)^{(2-\alpha)(k+1)} + \sum_{k=0}^{\infty} c_{2k+1} (t_s - t)^{(2-\alpha)(k+1)+1} , \quad (11.13)$$

where $H_s = \text{const} > 0$, $\psi_s = a_s^{1/n} = \text{const} > 0$ and

$$c_{2k} = \frac{(\kappa u_s^2)^{k+1} \psi_s}{(2-\alpha)^{k+1}(k+1)! \prod_{m=1}^{k+1} [m(2-\alpha) - 1]},$$

$$c_{2k+1} = -\frac{(\kappa u_s^2)^{k+1} n H_s \psi_s}{(2-\alpha)^{k+1}(k+1)! \prod_{m=1}^{k+1} [m(2-\alpha) + 1]}.$$

Using (11.11) we get

$$H(t_s) = H_s, \quad \left(\frac{\ddot{a}}{a}\right)_{t \to t_s} = \frac{\kappa u_s^2}{n(t_s - t)^{\alpha}} - (n-1)H_s^2,$$

so the solution (11.13) contains the SFS at $t \to t_s$ for $\alpha < 1$.

At last, let's consider the (11.7). After differentiation one gets

$$\frac{d^m \psi}{dt^m} = (-1)^m \sum_{k=1}^{\infty} c_k \prod_{l=0}^{m-1} [k(2-\alpha) - l] (t_s - t)^{k(2-\alpha)-m} , \quad (11.14)$$

so we have a singularity in (11.14) for

$$k < \frac{m}{2-\alpha}. \quad (11.15)$$

Since $0 < \alpha < 1$ then

$$\frac{1}{2} < \frac{1}{2-\alpha} < 1.$$

Therefore for $m > 1$ the expression (11.14) diverge as the cosmological singularity is approached. Thus we have obtained the solution which contains a some kind of generalization of the Big Break singularity (the scale factor remains finite and the Hubble parameter vanishes as singularity is approached). In the case of positive curvature one has to choose $n = 1$ (cf. Remark 11.5) and we have a "brane-like" singularity (the density is finite and positive whereas all higher derivatives of the scale factor, starting out from the second one, diverge as the cosmological singularity is approached). In the next section we'll present a couple of models with the self-acting and minimally coupled scalar fields or with the homogeneous tachyon fields $T = T(t)$ described by Sens or Born-Infeld type Lagrangians which result in such behavior.

The similar investigation can be done for the (11.13). This singularity is characterized by the fact that Hubble parameter remains finite instead of vanishing as the cosmological singularity is approached. This solution describes the appearance of a "brane-like" singularity in the flat universe.

11.4   Field models

If the universe is filled with a self-acting and minimally coupled scalar field with Lagrangian (11.5) then the energy density and pressure are

$$\rho = K + V, \quad p = K - V,$$

therefore

$$V = \frac{1}{2}(\rho - p), \quad K = \frac{1}{2}(\rho + p). \tag{11.16}$$

Using (11.3) one can write

$$K = \frac{\dot{\psi}^2 - \psi\ddot{\psi}}{3n\psi^2} = \frac{(w+1)\dot{\psi}^2}{2n^2\psi^2}, \tag{11.17}$$

$$V = \frac{n\psi\ddot{\psi} + (3-n)\dot{\psi}^2}{3n^2\psi^2} = \frac{(1-w)\dot{\psi}^2}{2n^2\psi^2}, \tag{11.18}$$

where

$$w = \frac{p}{\rho} = -1 + \frac{2n}{3}\left(1 - \frac{\ddot{\psi}\psi}{\dot{\psi}^2}\right). \tag{11.19}$$

The second model is the universe filled with a homogeneous *tachyon field* $T = T(t)$ described by the *Sens* Lagrangian density [174]:

$$L = -V(T)\sqrt{1 - g_{00}\dot{T}^2}, \tag{11.20}$$

and in a flat Friedmann universe with metric $ds^2 = dt^2 - a^2(t)d\Omega^2$ we have density and pressure

$$\rho = \frac{V(T)}{\sqrt{1 - \dot{T}^2}}, \quad p = L, \tag{11.21}$$

so

$$\dot{T}^2 = 1 + w, \quad V = \frac{\dot{\psi}^2}{n^2\psi^2}\sqrt{-w}, \tag{11.22}$$

where $w = p/\rho$ is defined by the (11.19).

The expression for V (11.22) holds iff $w < 0$. If $w > 0$ one should introduce a new field theory based on a Born-Infeld type action with Lagrangian [163]

$$L = W(T)\sqrt{\dot{T}^2 - 1}, \tag{11.23}$$

so

$$\rho = \frac{W(T)}{\sqrt{\dot{T}^2 - 1}}, \quad p = L,$$

and

$$\dot{T}^2 = 1 + w, \quad W = \frac{\dot{\psi}^2}{n^2\psi^2}\sqrt{w}. \tag{11.24}$$

It is interesting that models (11.20) and (11.23) can be connected via the so called "transgression of the boundaries" [163].

Now, using the potential (11.6) and the solution (11.7) we get

$$w = -\frac{2n(\alpha - 1)^2}{3\kappa u_s^2(t_s - t)^{2-\alpha}}. \tag{11.25}$$

Therefore, for the case $\kappa = +1$ we have $w < 0$ (BFS) and $w \to -\infty$ as the cosmological singularity is approached. Using (11.17) and (11.22) one concludes that this will be the case for the phantom (both usual ϕ and tachyon T) fields only — the case which we leave out of consideration in this section.

For the case $\kappa = -1$ and $0 < \alpha < 1$ one has a Big Break singularity with $w \to +\infty$. For the model (11.5) we get

$$\dot{\phi}^2 = \frac{2u_s^2}{3n(t_s - t)^\alpha} > 0,$$

and

$$p \to \frac{2u_s^2}{3n(t_s - t)^\alpha} \to +\infty, \quad H \to 0,$$

as the cosmological singularity is approached. So all energy conditions are satisfied and $H_s = 0$ as it should be for a BBS. If $k = +1$ we have a "brane-like" singularity with

$$\rho_s = \frac{u_s^4}{(2 - \alpha)^2 (1 - \alpha)^2}. \tag{11.26}$$

At $t \to t_s$ the potential $V = V(\phi)$ and field $\phi = \phi(t)$ are given by expressions

$$V(\phi) = -\frac{u_s^2}{3n} \left(\frac{3n(2 - \alpha)}{8u_s} \right)^{-\alpha/(2-\alpha)} (\phi - \phi_s)^{-2\alpha/(2-\alpha)}, \tag{11.27}$$

$$\phi = \phi_s \mp \frac{2u_s}{2 - \alpha} \sqrt{\frac{2}{3n}} (t_s - t)^{1-\alpha/2},$$

and $V(\phi) \to -\infty$, $\phi \to \phi_s$ at $t \to t_s$.

For the solution (11.13) we have the same potential (11.27) and

$$w = -\frac{2\kappa u_s^2}{3nH_s^2(t_s - t)^\alpha},$$

instead of (11.25).

In the case of *tachyon cosmology* one should use the model (11.23). It easy to see that $\dot{T}^2 > 0$ and

$$T(t) = T_s \mp 2\sqrt{\frac{2n}{3} \frac{1 - \alpha}{\alpha u_s}} (t_s - t)^{\alpha/2} \to T_s,$$

$$W(\Phi) = \sqrt{\frac{2}{3n}} \frac{u_s^3}{n(1 - \alpha)} \Phi^{2/\alpha - 3} \to 0, \tag{11.28}$$

with

$$\Phi = \frac{1}{2} \sqrt{\frac{3}{2n} \frac{\alpha u_s}{1 - \alpha}} (T - T_s).$$

Since $0 < \alpha < 1$ then $2/\alpha - 3 > -1$. For example, if $\alpha = 2/5$ then

$$W(T) = \frac{5\sqrt{6} u_s^5}{54 n^{5/2}} (T - T_s)^2,$$

and for the $\alpha = 2/7$

$$W(T) = \frac{21\sqrt{6} u_s^7}{12500 n^{7/2}} (T - T_s)^4.$$

Finally, let us consider the equation of state which results in the potential (11.27). It was shown in [163] that the equation of state

$$\rho + p = \gamma \rho^\lambda, \tag{11.29}$$

results in dynamics which might be described by the self-acting potential is the form

$$V(\phi) = Q^{-2/(\lambda-1)} - \frac{\gamma}{2} Q^{-2\lambda/(\lambda-1)}, \tag{11.30}$$

with

$$Q = \frac{3\sqrt{\gamma}(\lambda-1)(\phi - \phi_s)}{2},$$

where $\gamma > 0$, $\lambda > 1$. When $\phi \to \phi_s$ we have

$$V(\phi) \to -\frac{\gamma}{2} Q^{-2\lambda/(\lambda-1)} \to -\infty.$$

Unfortunately, this expression is just formally equivalent to potential (11.27). In fact, the second term in (11.30) (which is the dominant one at $\phi \to \phi_s$) is exactly (11.27) if $\alpha = 2\lambda/(2\lambda - 1)$, therefore for $\lambda > 1$ we get $1 < \alpha < 2$. It means that $\rho_s = \infty$ and we have a singularity of the III type which, in the case of general position, is not a "brane-like" one.

Near the singularity, the correct equation of state for the solution (11.7) in case of a positive curvature has the form which looks similar to (11.29):

$$\rho + 3p = \gamma (\rho_s - \rho)^{-|\lambda|}, \tag{11.31}$$

where

$$|\lambda| = \frac{\alpha}{2(1-\alpha)},$$

and $0 < \alpha < 1$.

11.5   The tachyon potentials

The method of linearization proves to be extremely useful in finding the potentials of the exact solvable *tachyon* models. In particular, as we shall see, the tachyon model which was discussed in detail in [163] is one of the simplest models in framework of the method of linearization.

Let's consider Eq. (11.3) with potential $U = 0$. The solution of this equation $\psi = Ct + C' \to Ct$ by the translation $t \to t - C'/C$. In this case we get

$$w = \frac{p}{\rho} = -1 + \frac{2n}{3}, \quad p = \frac{2n-3}{3n^2 t^2}, \quad \dot{T}^2 = w + 1,$$

so

$$T = \pm \sqrt{\frac{2n}{3}} (t - t_s) + T_s,$$

and

$$V(T) = \frac{2\sqrt{9-6n}}{n \left[2(T - T_s) \pm \sqrt{6n} t_s \right]^2}. \tag{11.32}$$

If $n = 3(1+k)/2$ with $-1 < k < +1$ (so $0 < n < 3$) and $t_s = 0$ then (11.32) has exactly the form of one of the potentials from the first paper [163] (with $T_0 \to T_s$).

The case with $U = \mu^2 = \text{const} > 0$ is a more interesting example. The solution of the (11.3) with Big Bang singularity at $t = 0$ is $\psi = C \sinh(\mu t)$. So

$$w = -1 + \frac{2n}{3 \cosh^2 \mu t}, \quad p = \frac{\mu^2 (2n - 3 \cosh^2 \mu t)}{2n^2 \sinh^2 \mu t}.$$

Using (11.22) one gets

$$T(t) = \pm \sqrt{\frac{2n}{3\mu^2}} \arctan(\sinh \mu t) + T_0,$$

and

$$V(t) = \frac{\mu^2 \cosh \mu t}{3n^2 \sinh^2 \mu t} \sqrt{9 \cosh^2 \mu t - 6n}.$$

Introducing $\Lambda = \mu^2/n^2$, $k = 2n/3 - 1$ we get

$$V(T) = \frac{\Lambda}{\sin^2 \xi} \sqrt{1 - (1+k) \cos^2 \xi}, \tag{11.33}$$

where

$$\xi = \frac{3}{2} \sqrt{\Lambda(1+k)} T.$$

(11.33) is the basic tachyon model of the paper [163].

It is not difficult to construct many other integrable tachyon models using the simple, solvable potentials of (11.3). Another fruitful way of doing it lies in a use of the *Darboux transformation* to those initial potentials $(U = 0, U = \mu^2)$.

11.6   The generalized singularities of type IV

The singularities of type IV (according to the classification of Ref. [168]) have the following behavior: for $t \to t_s$, $a \to a_s$, $\rho \to 0$, $|p| \to 0$ and the higher derivatives of H diverge $(0 < a_s < \infty)$. In this section we present a new type of singularities:

Generalized type IV: For $t \to t_s$, $a \to a_s$, $\rho \to \rho_s$, $p \to p_s$ and higher derivatives (starting out from the third one) of H diverge and $0 < a_s < \infty$, $0 < \rho_s < \infty$, $0 < |p_s| < \infty$.

Let's put $n = 3$, $\psi = a^3$ and (in parametric form)

$$\psi = A + \frac{\kappa B}{4} \left(4 \cos \eta - 2 \cos^2 \eta - \cos^4 \eta - 4 \log(1 + \cos \eta) \right),$$

$$t = t_s + \frac{1}{\kappa} \left(\log \left| \tan \frac{\eta}{2} \right| + \cos \eta \right), \tag{11.34}$$

where A, B, κ are constants, $0 \leq \eta \leq \pi$; $\eta = 0$ corresponds to $t = -\infty$, $\eta = \pi/2$ to $t = t_s$ (singularity) and $\eta = \pi$ to $t = +\infty$. After the differentiation we get (a dot denotes the derivative with respect to cosmic time t rather than to parameter η)

$$\dot{\psi} = \kappa^2 B \left(1 - \cos^3 \eta \right),$$

$$\ddot{\psi} = 3\kappa^3 B \sin^2 \eta,$$

$$\dddot{\psi} = \frac{6B\kappa^4 \sin^2 \eta}{\cos \eta},$$

$$\ddddot{\psi} = \frac{6\kappa^5 B \sin^2 \eta (\cos^2 \eta + 1)}{\cos^4 \eta},$$

and so on.

For $\kappa B > 0$ the function (11.34) is the monotonously increasing one for $0 \leq \eta \leq \pi$ and $\psi(\pi) = +\infty$ (i.e. at $t = +\infty$). Thus at $t = -\infty$

$$\psi = A + B\kappa \left(\frac{1}{4} - \log 2 \right), \quad \dot{\psi} = 0, \quad \ddot{\psi} = 0.$$

At $t = t_s$

$$\psi = A, \quad \dot{\psi} = \kappa^2 B, \quad \ddot{\psi} = 3\kappa^3 B$$

and, starting out from the third one, all higher derivatives diverge. At $t = +\infty$

$$\psi = \text{sign}(\kappa B) \times \infty, \quad \dot{\psi} = 2\kappa^2 B, \quad \ddot{\psi} = 0.$$

Therefore at $t = t_s$

$$\rho_s = \frac{\kappa^4 B^2}{9A^2}, \quad p_s = \frac{B\kappa^3(\kappa B - 6A)}{9A^2}, \quad w_s = 1 - \frac{6A}{\kappa B}.$$

Thus we have a generalization of a type IV singularity at t_s, where the density and pressure are finite and nonzero whereas all higher derivatives of H diverge.

It is convenient to introduce a new parameter s:

$$s = \frac{6A}{\kappa B},$$

so

$$\rho_s = \frac{4\kappa^2}{s^2}, \quad p_s = \frac{4\kappa^2(1-s)}{s^2}, \quad w_s = 1 - s.$$

If $4/3 < s < 2$ then $-1 < w_s < -1/3$; if $s = 2$ then $w_s = -1$; if $s > 2$ then $w_s < -1$. To obtain the initial Big Bang singularity at $t = t_i$, $-\infty < t_i < t_s$ one should put

$$s < s_i = 6\log 2 - \frac{3}{2} \sim 2.659,$$

or

$$w_s > w_i = -1.659.$$

In the initial Big Bang singularity, the *barotropic index* $w = +\infty$, on the other hand, $w(\eta)$ is the monotonously decreasing function. These properties result in a following conclusion: if

$$\frac{4}{3} < s < s_i,$$

then after Big Bang the model (11.34) will go through the usual expansion with damping, but starting out from some moment it will experience an accelerated expansion up to a future generalized type IV singularity.

Let us now briefly sum up what we have gathered so far. In this section we have discussed a simple method of construction of exact solutions of the Friedmann equations with finite scale factor singularities. Despite apparent simplicity, the method allows for acquirement of solutions characterized by the extremely interesting properties.

We can break the main results of this section into five parts:

(i) we have obtained a new type of finite-time, future singularities which seem to be most similar to the type IV of [168] but are different nevertheless as they have nonzero pressure and density at the singular point;

(ii) we have obtained a new type of finite-time, future quasi-singularities being rather similar to the w-singularities (which are quasi-singularities too) but having nonzero pressure at a singular point;

(iii) we have shown that "brane-like" singularities can occur in a common Friedmann cosmology with potential (11.27) and equation of state (11.31) (near of singularity) as well as in a tachyon cosmology with potential (11.28);

(iv) we have obtained the generalized Big Break singularities not only for the universe filled with tachyons but also a usual minimally coupled scalar field;

(v) we have shown that basic tachyon model which was discussed in detail in [163] is one of the simplest models in framework of the method of linearization.

Chapter 12

The exact cosmologies on the brane

12.1 Introduction

Brane world scenario have been proposed more then a decade ago and this scenario attracted a lot of attention. One reasons for such attention was the hope to understand and successfully describe the observed accelerated expansion of the Universe.

To our knowledge the work by Binetrruy, Deffaet and Langlois [175] was the first work where brane cosmology, different from a standard Friedmann cosmology, have been proposed. It is interesting to note that the first work on the brane cosmology was in a huge part dedicated to a search for the exact solutions. This situation is a diametrically opposite to that in cosmological inflation theory where for about a decade only the approximated solution have been a rage of the season, up until the works [11, 16]. Further development of the brane cosmology scenarios was performed in works [176–180] where few exact solutions have been found. In the work [181] the relationship was shown between the exact solutions in [179, 180] and [176, 177]. Namely, the explicit coordinate transformation was given which proved the equivalence between these two solutions, i.e. that both solutions represent the same spacetime in different coordinate systems.

For consideration of the inflation scenario in brane cosmology it is important to analyze the scalar field solutions with self-interacting potential. There is a huge amount of works devoted to this issue. We will mention few of them where investigation of exact solutions was carried out. In the article [182] was found the exact solution at a high energy limit when H^2 proportional to ρ^2; canonical and tachyon fields were analyzed there. A general thick brane with a scalar field was analysed in the work [183], where restrictions on the scalar potential was obtained and the exact solutions for a stepwise potentials of different shapes was found. A general scalar field and barotropic fluid during the early stage of a brane-world where the Friedmann constraint is dominated by the square of the energy density was studied in the work [184].

In this chapter we present the method of exact solution construction and the new classes of exact solutions obtained via the superpotential method (see [71] and literature cited therein). Fist application of this method for Randal-Sandrum brane cosmology was perfomed in the work [185].

12.2   The method of construction of exact cosmological solutions on the brane

Let us consider the simplest brane model in which spacetime is homogeneous and isotropic along three spatial dimensions, being our 4-dimensional universe an infinitesimally thin wall, with constant spatial curvature, embedded in a 5-dimensional spacetime ([186], [187]). In the Gaussian normal coordinate system, for the brane which is located at $y = 0$, one gets

$$ds^2 = -n^2 dt^2 + a^2(t, y)\gamma_{ij} dx^i dx^j + dy^2, \qquad (12.1)$$

where γ_{ij} is the maximally 3-dimensional metric. Let t be the proper time on the brane ($y = 0$), then $n(t, 0) = 1$. Therefore, one gets the FRW metric on the brane

$$ds^2_{|y=0} = -dt^2 + a^2(t, 0)\gamma_{ij} dx^i dx^j. \qquad (12.2)$$

The 5-dimensional Einstein equations have the form

$$R_{AB} - \frac{1}{2} g_{AB} R = \chi^2 T_{AB} + \Lambda_4 g_{AB}, \qquad (12.3)$$

where Λ_4 is the bulk cosmological constant, $\chi^2 = 8\pi G^{(5)}/c^4$, $G^{(5)}$ is the gravitational constant in 5-dimensional spacetime. The next step is to write the total energy momentum tensor T_{AB} on the brane as

$$T_B^A = S_B^A \delta(y), \qquad (12.4)$$

with $S_B^A = \mathrm{diag}(-\rho_b, p_b, p_b, p_b, 0)$, where ρ_b and p_b are the total brane energy density and pressure, respectively.

One can now calculate the components of the 5-dimensional Einstein tensor which solves Einstein's equations. The crucial idea here is the usage of an appropriate junction conditions near $y = 0$. These can be reduced to the following two relations:

$$\frac{1}{n}\frac{dn}{dy}\Big|_{y=0+} = \frac{\chi^2}{3}\rho_b + \frac{\chi^2}{2}p_b, \qquad \frac{1}{a}\frac{da}{dy}\Big|_{y=0+} = -\frac{\chi^2}{6}\rho_b. \tag{12.5}$$

After some calculations, one obtains the following result

$$H^2 = \chi^4\frac{\rho_b^2}{36} + \frac{\Lambda_4}{6} - \frac{k}{a^2} + \frac{\mathcal{E}}{a^4}. \tag{12.6}$$

This expression is valid on the brane only. Here $H = \dot{a}(t,0)/a(t,0)$ and $\mathcal{E}$ is an arbitrary integration constant. The energy conservation equation is correct, too,

$$\dot{\rho}_b + 3\frac{\dot{a}}{a}(\rho_b + p_b) = 0. \tag{12.7}$$

Now, let $\rho_b = \rho + \lambda_b$, where λ_b is the brane tension. Further we consider the fine-tuned brane with $\Lambda_4 = \lambda_b^2\chi^4/6$ and the case of flat spacetime ($k = 0$):

$$\frac{\dot{a}^2}{a^2} = \frac{\lambda_b\chi^4}{6}\frac{\rho}{3}\left(1 + \frac{\rho}{2\lambda_b}\right) + \frac{\mathcal{E}}{a^4}. \tag{12.8}$$

In what follows we will consider a single brane model which mimics GR at present but differs from it at late times. We set $M_p^{-2} = 8\pi G = \sigma\chi^4/6$. For simplicity, we set $\mathcal{E} = 0$ (the term with $\mathcal{E}$ is usually called *"dark radiation"*). In fact, setting $\mathcal{E} \neq 0$ does not lead to additional solutions on a radically new basis, in the framework of our approach. Eq. (12.8) can be simplified to

$$\frac{\dot{a}^2}{a^2} = \frac{\rho}{3M_p^2}\left(1 + \frac{\rho}{2\lambda_b}\right). \tag{12.9}$$

One can see that Eq. (12.9), for $\rho << |\lambda_b|$, differs insignificantly from the FRW equation. The brane model with a positive tension has been discussed in [188]–[191] in the context of the unification of early- and late-time acceleration eras. The braneworld model with a negative tension and a time-like extra dimension can be regarded as being dual to the Randall-Sundrum model ([137], [192], [193]). Note that, for this model, the Big Bang singularity is absent. And this fact does not depend upon whether or not matter violates the energy conditions ([194]). This same scenario has also been used to construct cyclic models for the universe [195].

One can assume that in our epoch the $\rho/2\lambda << 1$ and so there is no significant difference between the brane model and FRW cosmology. But the universe's evolution in the future or in past, for brane cosmology, can in fact differ from such convenient cosmology, due to the non-linear dependence of the expansion rate on the energy density.

One can reduce the field equation to the *slow-roll form*

$$3HU = -W'_\phi, \tag{12.10}$$

with substitution

$$W = V + 1/2U^2, \quad U(\phi) = \dot\phi \tag{12.11}$$

Then Friedmann equation (12.9) can be rewritten in terms of the *superpotential W*

$$H^2 = \frac{1}{3M_p^2} W \left(1 + \frac{W}{2\lambda_b} \right). \tag{12.12}$$

Considering H as positive and inserting (12.12) into (12.10) one can obtain

$$\frac{\sqrt3}{M_p} U(\phi) \sqrt{W \left(1 + \frac{W}{2\lambda_b} \right)} = -W'_\phi. \tag{12.13}$$

This is the key equation for further progress. For given $W(\phi)$ as function of scalar field one can define the dependence of scalar field from time inversing the following relation:

$$t - t_0 = -\frac{\sqrt3}{M_p} \int \frac{d\phi}{W'_\phi} \sqrt{W \left(1 + \frac{W}{2\lambda_b} \right)} \tag{12.14}$$

In frames of this approach the physical potential is

$$V(\phi) = W(\phi) - \frac{M_p^2}{6} W'^2_\phi \left(W(\phi) \left(1 + \frac{W(\phi)}{2\lambda_b} \right) \right)^{-1} \tag{12.15}$$

One can also define $U(\phi)$ as function of scalar field. In this case the integration (12.13) via separation of W and ϕ leads to the superpotential and Hubble parameter presentation in quadratures

$$\sqrt W = \sqrt{2\lambda_b} \sinh \left(-\sqrt{\frac{3}{2\lambda_b}} \frac{1}{2M_P} \int U(\phi) d\phi \right), \quad W > 0, \tag{12.16}$$

$$\sqrt{-W} = \sqrt{2\lambda_b} \cosh \left(-\sqrt{\frac{3}{2\lambda_b}} \frac{1}{2M_P} \int U(\phi) d\phi \right), \quad W < 0, \tag{12.17}$$

$$H = \sqrt{\frac{\lambda_b}{6}} \frac{1}{M_P} \sinh \left(-\sqrt{\frac{3}{2\lambda_b}} \frac{1}{M_P} \int U(\phi) d\phi \right) \tag{12.18}$$

Note that the argument in (12.18) should be positive. Therefore $\int U(\phi) d\phi$ should be always negative.

The physical potential V can be obtained from the superpotential definition (12.11)

$$V(\phi) = 2\lambda_b \sinh^2 \left(-\sqrt{\frac{3}{2\lambda_b}} \frac{1}{2M_P} \int U(\phi) d\phi \right) - \frac{1}{2} U(\phi)^2. \qquad (12.19)$$

Thus to obtain the examples of exact solutions one can suggest the functional dependence a scalar field ϕ on cosmic time t and evaluate the integral $\int U(\phi) d\phi$:

$$\int U(\phi) d\phi = \int U^2(t) dt.$$

Standard calculation leads to the following formula for $N(\phi)$:

$$N(\phi) = -\sqrt{\frac{\lambda_b}{2}} M_P^{-2} \int \sqrt{W \left(1 + \frac{W}{2\lambda_b} \right)} \sqrt{\left(\frac{W}{\lambda_b} + 1 \right)^2 + 1} \frac{d\phi}{W'} \qquad (12.20)$$

To understand the period with accelerating Universe expansion let us calculate $\frac{\ddot{a}}{a} = \dot{H} + H^2$. For given $W(\phi)$ we have the following simple relation for acceleration parameter:

$$\frac{\ddot{a}}{a} = -\frac{1}{6} \frac{W'^2_\phi}{W} + \frac{W}{3M_P^2} + \frac{W}{6\lambda_b} \left(\frac{W}{M_P^2} - \frac{1}{2} \frac{W'^2_\phi}{W(1 + W/2\lambda_b)} \right) \qquad (12.21)$$

The first two terms corresponds to the case of usual FRW cosmology. The last appears due to the brane tension. Combining (12.21) with (12.14) one can analyze the behavior $\ddot{a}/a$ as function of time.

If we define the function $U(\phi)$ it is convenient to use the result

$$\frac{\ddot{a}}{a} = -\frac{U^2}{2M_P^2} \cosh \left(-\sqrt{\frac{3}{2\lambda_b}} \frac{1}{M_P} \int U(\phi) d\phi \right)$$

$$+ \frac{\lambda_b}{6M_P^2} \sinh^2 \left(-\sqrt{\frac{3}{2\lambda_b}} \frac{1}{M_P} \int U(\phi) d\phi \right) \qquad (12.22)$$

To investigate changing of the acceleration sign let us represent (12.22) in the following form

$$\frac{6M_P^2}{\lambda_b} \frac{\ddot{a}}{a} = Z^2 - \frac{3U^2}{\lambda_b} Z - 1, \quad Z = \cosh \left(-\sqrt{\frac{3}{2\lambda_b}} \frac{1}{M_P} \int U(\phi) d\phi \right) \qquad (12.23)$$

Taking into account that $Z > 1$ we omit the the root

$$Z_2 = \frac{3U^2}{2\lambda_b} - \sqrt{\frac{9U^4}{4\lambda_b^2} + 1} \qquad (12.24)$$

as it is less then unity. For the next root

$$Z_1 = \frac{3U^2}{2\lambda_b} + \sqrt{\frac{9U^4}{4\lambda_b^2} + 1} \qquad (12.25)$$

it is easy to check that $Z_1 > 1$ for any value of U^2. Therefore we can imply that there is only one point during the evolution when deceleration have been changed to acceleration. So in the framework of scalar field cosmology on the brane we have only one inflationary period. If it is an early inflation then once again as in FRW cosmology we have to verify the existence of exit from the inflation. On the other hand, if it is the later inflation, we will have to introduce at least one additional term to our equation which will be responsible for the early inflation at the primordial stage of Universe's evolution.

12.3   The solutions for a given superpotential

Let's consider two simple superpotentials. The first case is

$$W = \frac{m^2 \phi^2}{2} \qquad (12.26)$$

One can easy derive the dependence $t(\phi)$. The result is

$$t - t_0 = -\frac{\sqrt{6\lambda_b}}{2m^2} M_P^{-1} \left[\operatorname{arcsinh} \left(\frac{m\phi}{2\sqrt{\lambda_b}} \right) + \frac{m\phi}{2\sqrt{\lambda_b}} \sqrt{1 + \frac{m^2 \phi^2}{4\lambda_b}} \right]$$

For simplicity we put $\phi(t_0) = 0$. At $t \to \infty$ scalar field $\phi \to -\infty$. One can consider the moment $t < t_0$ for this moment $\phi > 0$. The universe acceleration is

$$\frac{\ddot{a}}{a} = -\frac{m^2}{12} + \frac{m^2 \phi^2}{6M_P^2} + \frac{m^2 \phi^2}{12\lambda_b} \left(\frac{m^2 \phi^2}{2M_P^2} - \frac{1}{2} \frac{m^2}{1 + m^2 \phi^2 / 4\lambda_b} \right) \qquad (12.27)$$

For scale factor as function of scalar field we have

$$a = a_0 \exp \left(\pm \frac{\phi^2}{8M_P^2} \left(2 + \frac{m^2 \phi^2}{4\lambda_b} \right) \right) \qquad (12.28)$$

One can choose the sign $-$ for $t < 0$ (i.e. $\phi > 0$) and $+$ for $t > 0$ ($\phi < 0$). The moment $t = -\infty$ corresponds to Big Bang singularity after which the universe expands with $\ddot{a} > 0$ before the non-inflationary phase follows.

Then universe again expands with acceleration. The asymptotic of solution is

$$a(t) \sim a_0 \exp\left(\frac{\lambda_b m^2}{2M_P^2}(t - t_0)^2\right). \tag{12.29}$$

The potential of scalar field can be obtained from (12.15):

$$V(\phi) = \frac{m^2\phi^2}{2} - \frac{M_P^2 m^2}{3}\left(1 + \frac{m^2\phi^2}{4\lambda_b}\right)^{-1}. \tag{12.30}$$

At $t \to \infty$ this potential corresponds to a free scalar field with mass m.

For the case of ϕ^4 superpotential

$$W(\phi) = \frac{\lambda\phi^4}{4} \tag{12.31}$$

we have the following link between time and scalar field:

$$t - t_0 = -\frac{\sqrt{3}}{2M_P\sqrt{\lambda}}\left(\cosh\eta - \cosh\eta_0 + \ln\tanh\frac{\eta}{2} - \ln\tanh\frac{\eta_0}{2}\right), \tag{12.32}$$

$$\eta = \text{arcsinh}\left(\left(\frac{\lambda}{8\lambda_b}\right)^{1/2}\phi^2\right).$$

The scalar field at the moment $t = t_0$ is $\phi_0 = \left(\frac{8\lambda_b}{\lambda}\right)^{1/4}\sinh^{1/2}\eta_0$ where η_0 is arbitrary constant. At $t \to \infty$ the scalar field $\phi \to 0$. From expression for universe acceleration

$$\frac{\ddot{a}}{a} = -\frac{2\lambda}{3}\phi^2 + \frac{\lambda}{12M_P^2}\phi^4 + \frac{\lambda^2\phi^6}{24\lambda_b}\left(\frac{1}{4M_P^2}\phi^2 - \frac{2}{1 + \lambda\phi^4/8\lambda_b}\right) \tag{12.33}$$

one can see that exit from inflation occurs. The scale factor as function of scalar field

$$a = a_0 \exp\left(\pm\frac{\phi^2}{8M_P^2}\left(1 + \frac{\lambda\phi^4}{24\lambda_b}\right)\right) \tag{12.34}$$

The negative sign corresponds to universe starting from $a = 0$ ($\phi = \infty$ at $t = -\infty$) and expanding with acceleration prior to some moment of time when exit from inflation occurs.

12.4 The solutions for a given scalar field as a function of time

The shape the exact solutions will take will naturally depend upon the particulars dynamics of the scalar fields. There is a lot to choose from, but for the didactic purposes we will start with the simplest cases.

12.4.1 *Logarithmic evolution of the scalar field*

Let

$$\phi = A \ln(\lambda t).$$

The solutions are

$$\sqrt{W} = \sqrt{2\lambda_b} \sinh\left(-\sqrt{\frac{3}{8\lambda_b}}\frac{1}{2M_P}\left(C_1 - \frac{A^2}{t}\right)\right), \quad W > 0, \qquad (12.35)$$

$$\sqrt{-W} = \sqrt{2\lambda_b} \cosh\left(-\sqrt{\frac{3}{2\lambda_b}}\frac{1}{2M_P}\left(C_1 - \frac{A^2}{t}\right)\right), \quad W < 0, \qquad (12.36)$$

$$H = \sqrt{\frac{\lambda_b}{6}}\frac{1}{M_P}\sinh\left(-\sqrt{\frac{3}{2\lambda_b}}\frac{1}{M_P}\left(C_1 - \frac{A^2}{t}\right)\right). \qquad (12.37)$$

The superpotential's and physical potential's presentation in terms of scalar field can be obtained using given dependance $\phi = A\ln(\lambda t)$.

$$\sqrt{W} = \sqrt{2\lambda_b} \sinh\left(\sqrt{\frac{3}{2\lambda_b}}\frac{1}{2M_P}\left(A^2\lambda\exp(-\frac{\phi}{A}) - C_1\right)\right), \quad W > 0, \tag{12.38}$$

$$\sqrt{-W} = \sqrt{2\lambda_b} \cosh\left(-\sqrt{\frac{3}{2\lambda_b}}\frac{1}{2M_P}\left(C_1 - A^2\lambda\exp(-\phi/A)\right)\right), \quad W < 0, \tag{12.39}$$

$$V(\phi) = 2\lambda_b \sinh^2\left(-\sqrt{\frac{3}{2\lambda_b}}\frac{1}{2M_P}\left(C_1 - A^2\lambda\exp(-\phi/A)\right)\right) - \frac{A^2\lambda^2}{2}e^{-2\phi/A}. \tag{12.40}$$

The potential $V(\phi)$ is depicted on Fig. 12.1. Thus we have obtained the exact formulas for given evolution of the scalar field. We know that for solutions under consideration we may have only one inflection point for scalar factor a associated with the Eq. (12.25). Using the definition for Z (12.23) one can obtain the equation for time corresponding to inflection point:

$$\frac{3A^2}{2\lambda_b t^2} + \sqrt{\frac{9A^4}{4\lambda_b^2 t^4} + 1} = \cosh\left(\sqrt{\frac{3}{2\lambda_b}}\frac{A^2}{M_P t}\right) \qquad (12.41)$$

It is easy to see that this equation will be true when $t \to \infty$. We can analyze this equation numerically to find the finite time $t < \infty$ which will

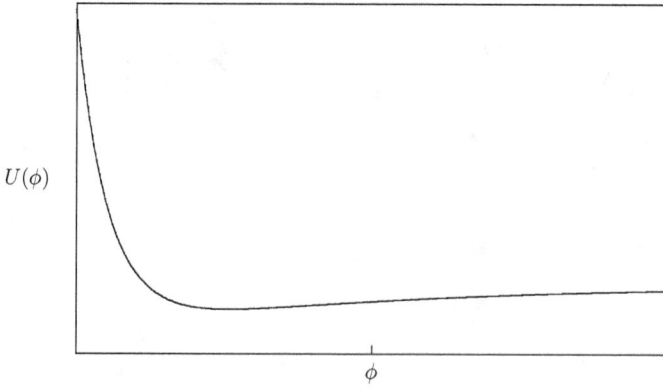

Fig. 12.1   The potential of scalar field for logarithmic evolution of scalar field.

correspond to deceleration changes to acceleration. One can state that this time is about $t_i \approx 0.286\lambda_b^{-1/2}$ in the units with $M_P = 1$ and for $A = 1$. Therefore we can use this approximate result to set this time equal to beginning of (early or later) inflation. From the other hand we can analyze the transition from brane cosmology to Friedmann one by tending the brane tension to infinity $\lambda_b \to \infty$. The results are:

$$\sqrt{\lambda_b} \to \infty :$$

$$\sqrt{W} = -\sqrt{3}\frac{1}{2M_p}\left(C_1 - \frac{A^2}{t}\right) = -\sqrt{3}\frac{1}{2M_p}\left(C_1 - A^2\lambda e^{-\phi/A}\right), \quad (12.42)$$

$$\sqrt{-W} = \infty, \quad (12.43)$$

$$H = -\frac{1}{\sqrt{6}M_p}\left(\sqrt{\frac{3}{2}}\frac{1}{M_p}\left(C_1 - \frac{A^2}{t}\right)\right), \quad (12.44)$$

$$V(\phi) = \infty. \quad (12.45)$$

Here we can define the scale factor with power law — exponential behavior

$$a(t) = \exp\left(-\frac{C_1 t}{2M_P^2}\right)t^{\frac{2A^2\lambda^2}{2M_P^2}}. \quad (12.46)$$

We carefully analyzed the solution in this section. To simplify presentation of the next solutions let us represent Eqs. (12.16)–(12.19) in the following way

$$\sqrt{W} = \sqrt{2\lambda_b} \sinh\left(-\sqrt{\frac{3}{2\lambda_b}} \frac{1}{2M_P} [F(\phi) + C_1]\right), \quad W > 0, \qquad (12.47)$$

$$\sqrt{-W} = \sqrt{2\lambda_b} \cosh\left(-\sqrt{\frac{3}{2\lambda_b}} \frac{1}{2M_P} [F(\phi) + C_1]\right), \quad W < 0, \qquad (12.48)$$

$$H = \sqrt{\frac{\lambda_b}{6}} \frac{1}{M_P} \sinh\left(-\sqrt{\frac{3}{2\lambda_b}} \frac{1}{M_P} [F(\phi(t)) + C_1]\right) \qquad (12.49)$$

$$V(\phi) = 2\lambda_b \sinh^2\left(-\sqrt{\frac{3}{2\lambda_b}} \frac{1}{2M_P} [F(\phi) + C_1]\right) - \frac{1}{2}[U(\phi)]^2. \qquad (12.50)$$

Using the general formulas (12.47)–(12.50) we will display new solutions by putting values for $F(\phi)$ and $U(\phi)$. Also we will use general formulas for transition to Friedmann Universe (by setting $\lambda_b \to \infty$) below

$$\sqrt{\lambda_b} \to \infty :$$

$$\sqrt{W} = -\frac{\sqrt{3}}{2M_p} (F(\phi) + C_1), \qquad (12.51)$$

$$\sqrt{-W} = \infty, \qquad (12.52)$$

$$H = -\frac{1}{2M_p^2} (F(\phi(t)) + C_1), \qquad (12.53)$$

$$V(\phi) = -\frac{3}{4M_P} (F(\phi) + C_1)^2 - \frac{1}{2}[U(\phi)]^2. \qquad (12.54)$$

12.4.2   *Power law evolution*

Let

$$\phi = At^s, \quad s \neq 0, \quad s \neq 1/2.$$

The solutions are represented by functions

$$F(\phi) = \frac{A^{-1/s}s^2}{2s-1} \phi^{2-1/s} = \frac{A^2 s^2}{2s-1} t^{2s-1}, \quad U^2(\phi) = A^2 s^2 \left[\frac{\phi}{A}\right]^{\frac{2s-2}{s}}. \qquad (12.55)$$

The equation for the time corresponding to inflection point is

$$\frac{3A^2 s^2}{2\lambda_b} t^{2s-2} + \sqrt{\frac{9A^4 s^4}{4\lambda_b} t^{4s-4} + 1} = \cosh\left(-\sqrt{\frac{3}{2\lambda_b}} \frac{A^2 s^2}{(2s-1)M_P} t^{2s-1}\right) \qquad (12.56)$$

By tending the brane tension to infinity $\lambda_b \to \infty$ we obtain

$$\sqrt{\lambda_b} \to \infty :$$

$$\sqrt{W} = -\frac{\sqrt{3}}{2M_p} \left(\frac{A^2 s^2}{2s-1} \left[\frac{\phi}{A} \right]^{\frac{2s-1}{s}} + C_1 \right), \qquad (12.57)$$

$$\sqrt{-W} = \infty, \qquad (12.58)$$

$$H = -\frac{1}{\sqrt{6}M_p} \left(\sqrt{\frac{3}{2}} \frac{1}{M_p} \left(C_1 + \frac{A^2 s^2}{2s-1} t^{\frac{2s-1}{s}} \right) \right), \qquad (12.59)$$

$$V(\phi) = \infty. \qquad (12.60)$$

In the case when $s = 1/2$ we have

$$F(\phi) = \frac{A^2}{2} \ln \frac{\phi}{A}, \quad U^2(\phi) = \frac{A^4}{4\phi^2} \qquad (12.61)$$

The potential of scalar field in this case is depicted on Fig. 12.2. The equation for the time corresponding to inflection point is

$$\frac{3A^2}{8t\lambda_b} + \sqrt{\frac{9A^4}{32t^2\lambda_b^2} + 1} = \frac{1}{2} \left(t^{-B} + t^B \right), \quad B = \sqrt{\frac{3}{2\lambda_b}} \frac{A^2}{4M_P} \qquad (12.62)$$

This time is about $t \approx 4.11\lambda_b^{-1/2}$ in units $M_P = A = 1$. In this moment the deceleration begins.

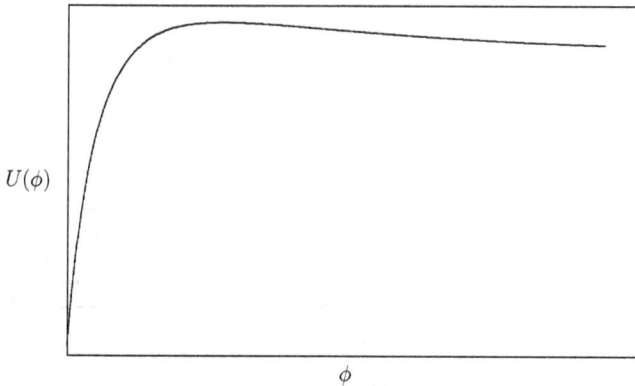

Fig. 12.2 The potential of scalar field for $\phi = At^{1/2}$.

12.4.3   *Exponential evolution*

Let

$$\phi = Ae^{-\lambda t}.$$

The solutions represented by functions

$$F(\phi) = -\frac{\phi^2 \lambda}{2} = -\frac{A^2 \lambda}{2} e^{-2\lambda t}, \quad U^2(\phi) = \lambda^2 \phi^2 \qquad (12.63)$$

The potential of scalar field is presented on Fig. 12.3.

The equation for the time corresponding to inflection point is

$$\frac{3A^2 \lambda^2}{2\lambda_b} e^{-2\lambda t} + \sqrt{\frac{9A^4 \lambda^4}{4\lambda_b^2} e^{-4\lambda t} + 1} = \cosh\left(\sqrt{\frac{3}{2\lambda_b}} \frac{A^2 \lambda}{M_P} e^{-2\lambda t}\right), \qquad (12.64)$$

For inflection time we have two different solutions: i) when $\lambda > 0$ $\ddot{a} \to 0$ at $t \to \infty$ (and $\ddot{a} > 0$ at $0 < t < \infty$); ii) when $\lambda = -\sqrt{\lambda_b} < 0$ $\ddot{a} = 0$ at $t \approx 0.879 \lambda_b^{-1/2}$ ($M_p = A = 1$). At this time acceleration changes to deceleration i.e. we have the inflation phase during finite time.

By tending the brane tension to infinity $\lambda_b \to \infty$ we obtain

$$\sqrt{\lambda_b} \to \infty : \qquad (12.65)$$

$$\sqrt{W} = -\sqrt{3} \frac{1}{2M_p} \left(C_1 - \frac{A^2 \lambda}{2} e^{-2\lambda t}\right), \qquad (12.66)$$

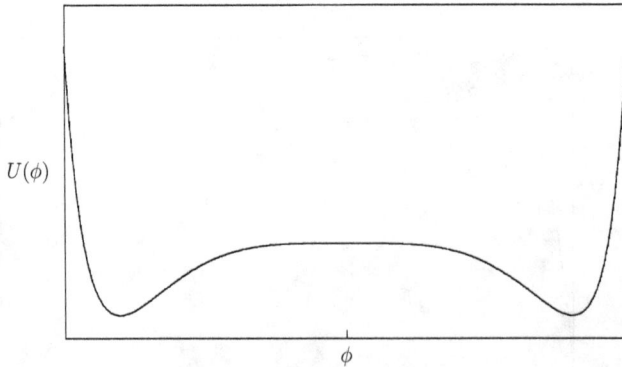

Fig. 12.3   The potential of scalar field for $\phi = A\exp(-\lambda t)$.

$$\sqrt{-W} = \infty, \qquad (12.67)$$

$$H = -\frac{1}{\sqrt{6}M_p}\left(\sqrt{\frac{3}{2}\frac{1}{M_p}\left(C_1 - \frac{A^2\lambda}{2}e^{-2\lambda t}\right)}\right), \qquad (12.68)$$

$$V(\phi) = \infty. \qquad (12.69)$$

The solutions above have been obtained earlier in [185] with a slightly different form. Here we presented them in more suitable way and gave detailed analysis. The solutions below are obtained first time and based on the scalar field evolutions considering in cosmology [26, 196]. Investigation of cosmological parameters for such evolution of scalar field was performed in the work [197].

12.4.4 *New classes of solutions*

12.4.4.1 $\phi = A\ln(\tanh(\lambda t))$

The solution is represented by functions

$$F(\phi) = A^2\lambda\left(2\cosh(\phi/A)\right) = A^2\lambda\left(\tanh(\lambda t) + \coth(\lambda t)\right), \qquad (12.70)$$

$$U^2(\phi) = \frac{A^2\lambda^2(1 - \exp(\frac{2\phi}{A}))^2}{\exp(\frac{2\phi}{A})} \qquad (12.71)$$

The potential of scalar field is depicted on Fig. 12.4.

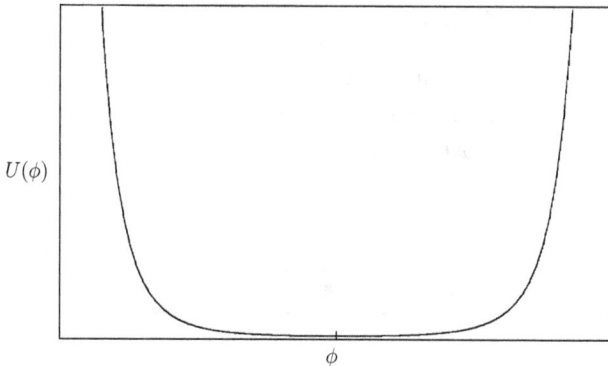

Fig. 12.4   The scalar field potential for $\phi = A\ln(\tanh(\lambda t))$.

The inflection point for the scale factor $a(t)$ can be obtained as a solution of the following equation:

$$\frac{3A^2\lambda^2}{2\lambda_b \cosh^2(\lambda t)\sinh^2(\lambda t)} + \sqrt{\frac{9A^4\lambda^4}{4\lambda_b^2 \cosh^4(\lambda t)\sinh^4(\lambda t)} + 1}$$

$$= \cosh\left(\sqrt{\frac{3}{2\lambda_b}\frac{A^2\lambda}{M_p}}(\tanh(\lambda t) + \coth(\lambda t))\right) \tag{12.72}$$

This equation has solutions at $\lambda < 0.82\lambda_b^{1/2}$ (for $A = M_P = 1$). The accelerated expansion begins at $t = t_1$ and ends at $t = t_2$. In Table 12.1 the duration of inflation stage are given for various λ.

12.4.4.2 $\phi = A\ln(\tan(\lambda t))$

The solution is represented by formulas

$$F(\phi) = A^2\lambda(2\sinh(\phi/A)) = A^2\lambda(\tan(\lambda t) - \cot(\lambda t)), \tag{12.73}$$

$$U^2(\phi) = \frac{A^2\lambda^2(1 + \exp(\frac{2\phi}{A}))^2}{\exp(\frac{2\phi}{A})} \tag{12.74}$$

On Fig. 12.5 one can see the dependence of potential from scalar field. The time for inflection point can be found as a solution of equation:

$$\frac{3A^2\lambda^2}{2\lambda_b \cos^2(\lambda t)\sin^2(\lambda t)} + \sqrt{\frac{9A^4\lambda^4}{4\lambda_b^2 \cos^4(\lambda t)\sin^4(\lambda t)} + 1}$$

$$= \cosh\left(\sqrt{\frac{3}{2\lambda_b}\frac{A^2\lambda}{M_p}}(\tan(\lambda t) - \cot(\lambda t))\right) \tag{12.75}$$

We have the same situation as in previous case. For various values of λ we have two roots corresponding to moment of beginning of inflation and to moment of exit from inflation. In Table 12.2 these results are given for various values of λ.

Table 12.1 The duration of accelerated expansion for model $\phi = A\ln(\tanh(\lambda t))$ for various λ.

$\lambda/\lambda^{1/2}$	
0.8	$0.38 < t\lambda_b^{1/2} < 0.53$
0.6	$0.315 < t\lambda_b^{1/2} < 0.382$
0.4	$0.295 < t\lambda_b^{1/2} < 1.2$
0.2	$0.286 < t\lambda_b^{1/2} < 2.31$

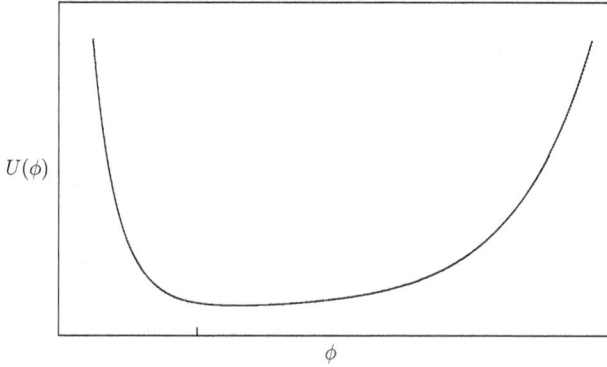

Fig. 12.5 The scalar field potential for $\phi = A\ln(\tan(\lambda t))$.

Table 12.2 The duration of accelerated expansion for model $\phi = A\ln(\tan(\lambda t))$ for various λ.

$\lambda/\lambda^{1/2}$	
0.6	$0.265 < t\lambda_b^{1/2} < 2.355$
0.8	$0.251 < t\lambda_b^{1/2} < 1.712$
1.0	$0.239 < t\lambda_b^{1/2} < 1.335$
1.2	$0.225 < t\lambda_b^{1/2} < 1.083$
1.4	$0.215 < t\lambda_b^{1/2} < 0.907$

12.4.4.3 $\phi = A/\sinh(\lambda t)$

The solution is described by the functions

$$F(\phi) = A^2\lambda\coth(\lambda t)\frac{1}{3}\left[\frac{\phi^2}{A^2} - 1\right] = A^2\lambda\coth(\lambda t)\frac{1}{3}\left[\frac{1}{\sinh^2(\lambda t)} - 1\right],$$

$$U^2(\phi) = \frac{\lambda^2}{A^2}\phi^4\left(1 + \left(\frac{A}{\phi}\right)^2\right) \tag{12.76}$$

The potential of scalar field is depicted on Fig. 12.6.
The inflection point can be found from corresponding equation

$$\frac{3A^2\lambda^2\cosh^2(\lambda t)}{2\lambda_b\sinh^4(\lambda t)} + \sqrt{\frac{9A^4\lambda^4\cosh^4(\lambda t)}{4\lambda_b^2\sinh^8(\lambda t)} + 1}$$

$$= \cosh\left(\sqrt{\frac{3}{2\lambda_b}}\frac{A^2\lambda\cosh(\lambda t)}{3M_p}\left[\frac{1}{\sinh^2(\lambda t)} - 1\right]\right) \tag{12.77}$$

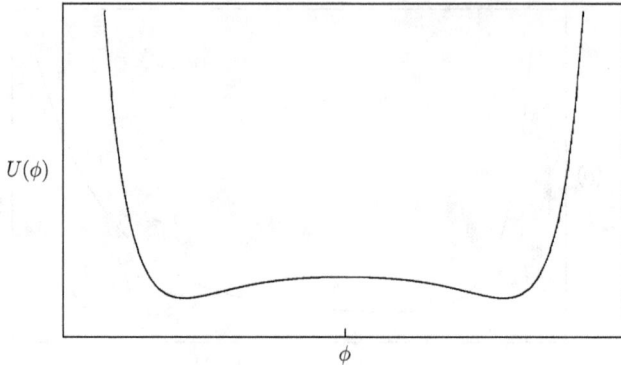

Fig. 12.6   The scalar field potential for $\phi = A/\sinh(\lambda t)$.

Table 12.3   The duration of accelerated expansion for model $\phi = A/\sinh(\lambda t)$ for various λ ($A = M_P = 1$).

$\lambda/\lambda^{1/2}$	
0.6	$0.665 < t\lambda_b^{1/2} < 3.5$
0.8	$0.545 < t\lambda_b^{1/2} < 2.63$
1.0	$0.468 < t\lambda_b^{1/2} < 2.11$
1.2	$0.414 < t\lambda_b^{1/2} < 1.765$
1.4	$0.374 < t\lambda_b^{1/2} < 1.52$

Once again for various values of λ we have phase of accelerated expansion during finite time. In Table 12.3 one can see results for various values of λ.

12.4.4.4 $\phi = A\arctan(\exp(\lambda t))$

The solution is represented by functions

$$F(\phi) = -\frac{A^2\lambda\cos^2(\phi/A)}{2} = -\frac{A^2\lambda}{2(1+\exp(2\lambda t))}, \qquad (12.78)$$

$$U^2(\phi) = \frac{A^2\lambda^2\tan^2(\frac{\phi}{A})}{(1+\tan^2(\frac{\phi}{A}))}. \qquad (12.79)$$

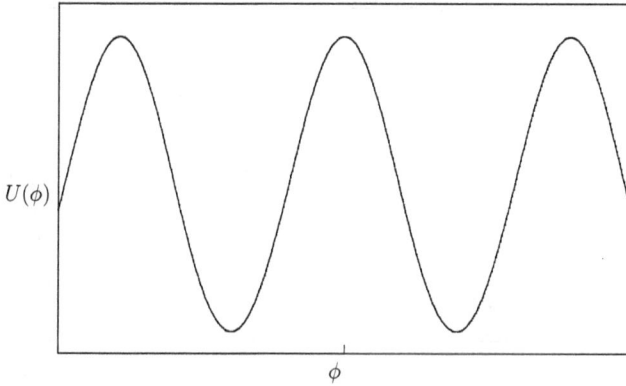

Fig. 12.7 The scalar field potential for $\phi = A \arctan(\exp(\lambda t))$.

The potential of scalar field is presented on Fig. 12.7.
From the equation

$$\frac{3A^2\lambda^2 \exp(2\lambda t)}{2\lambda_b(1 + \exp(2\lambda t))^2} + \sqrt{\frac{9A^4\lambda^4 \exp(4\lambda t)}{4\lambda_b^2(1 + \exp(2\lambda t))^4} + 1}$$

$$= \cosh\left(\sqrt{\frac{3}{2\lambda_b}}\frac{A^2\lambda}{2M_p(1 + \exp(2\lambda t))}\right) \tag{12.80}$$

one can find that for $\lambda > 0$ $\ddot{a} > 0$ $(0 < t < \infty)$ and for $\lambda < 0$ we have the moment when deceleration begins. For example for $\lambda = -\lambda_b^{1/2}$ we have $t \approx 0.093\lambda_b^{-1/2}$.

12.4.4.5 $\phi = A \sin^{-1}(\lambda t)$

The solution is described by formulas

$$F(\phi) = A^2\lambda \cot(\lambda t)\frac{1}{3}\left[1 - \frac{1}{\sin^2(\lambda t)}\right] = A^2\lambda \cot(\lambda t)\frac{1}{3}\left[1 - \frac{1}{\sin^2(\lambda t)}\right],$$
$$\tag{12.81}$$

$$U^2(\phi) = \frac{\lambda^2}{A^2}\phi^4\left(1 - \left(\frac{A}{\phi}\right)^2\right). \tag{12.82}$$

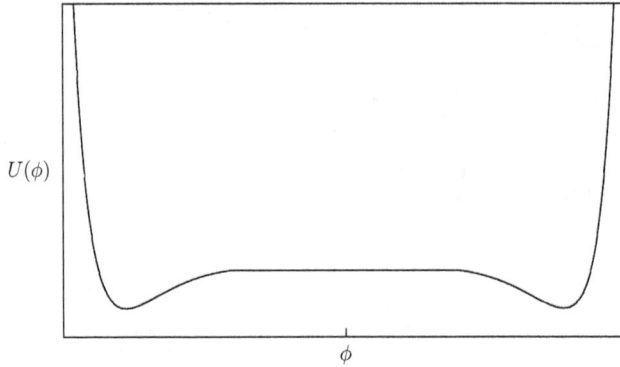

Fig. 12.8   The scalar field potential for $\phi = A \sin^{-1}(\lambda t)$.

On Fig. 12.8 one can see the potential of scalar field.
The inflection point can be found from equation

$$\frac{3A^2\lambda^2 \cos^2(\lambda t)}{2\lambda_b \sin^4(\lambda t)} + \sqrt{\frac{9A^4\lambda^4 \cos^4(\lambda t)}{4\lambda_b^2 \sin^8(\lambda t)} + 1}$$

$$= \cosh\left(\sqrt{\frac{3}{2\lambda_b}} \frac{A^2\lambda \cot(\lambda t)}{3M_p}\left[1 - \frac{1}{\sin^2(\lambda t)}\right]\right). \qquad (12.83)$$

For $\lambda = \lambda_b^{1/2}$ we have the phase of accelerated expansion for $t > 0.4\lambda_b^{-1/2}$.

Now would be a perfect time to stop and reflect on what we have managed to gain. We have discussed a simple method of construction of exact solutions for cosmological equations on RS brane. Despite simplicity, the method allows for acquirement of solutions characterized by interesting properties. These solutions have inflationary phases under quite general assumptions. This is an indication that inflationary regime seems a common occurrence in cosmology not requiring any special initial conditions. We have the following cases:

(i) accelerated expansion begins at some moment of time and lasts forever (logarithmic evolution of scalar field, $\phi \sim \sin^{-1}(\lambda t)$). In principle current observed acceleration can be described by this case.

(ii) acceleration take place before $t < t_f$ when deceleration starts (power and exponential evolution of scalar field or $\phi \sim \arctan(\exp(\lambda t))$). These solutions may correspond to inflation in beginning of cosmological evolution and exit from inflation phase.

(iii) acceleration occurs during interval $t_i < t < t_f$ ($\phi \sim \ln(\tanh(\lambda t))$, $\phi \sim \ln(\tan(\lambda t))$, $\phi \sim 1/\sinh(\lambda t)$). The interpretation of such solutions is obvious: we have the possible exit from inflation or current accelerated expansion or initial inflation rolling to later slow inflation (with further exit from it).

Therefore these models can not only describe inflation but also describe an exit from the inflationary phase without a fine tuning of the parameters. This fact can be seen as evidence of the existence of a realistic model that contains an inflationary phase in the early universe stage and a current phase of accelerated expansion.

Bibliography

[1] A. A. Starobinsky, A new type of isotropic cosmological models without singularity, *Phys. Lett.* **B91**, pp. 99–102 (1980), doi:10.1016/0370-2693(80)90670-X.

[2] A. H. Guth, The inflationary universe: A possible solution to the horizon and flatness problems, *Phys. Rev.* **D23**, pp. 347–356 (1981), doi:10.1103/PhysRevD.23.347.

[3] A. D. Linde, A new inflationary universe scenario: A possible solution of the horizon, flatness, homogeneity, isotropy and primordial monopole problems, *Phys. Lett.* **B108**, pp. 389–393 (1982), doi:10.1016/0370-2693(82)91219-9.

[4] A. Albrecht and P. J. Steinhardt, Cosmology for grand unified theories with radiatively induced symmetry breaking, *Phys. Rev. Lett.* **48**, pp. 1220–1223 (1982), doi:10.1103/PhysRevLett.48.1220.

[5] A. D. Linde, Chaotic inflation, *Phys. Lett.* **B129**, pp. 177–181 (1983), doi:10.1016/0370-2693(83)90837-7.

[6] V. A. Belinsky, I. M. Khalatnikov, L. P. Grishchuk and Ya. B. Zeldovich, Inflationary stages in cosmological models with a scalar field, *Phys. Lett.* **B155**, pp. 232–236 (1985), doi:10.1016/0370-2693(85)90644-6, [29(1985)].

[7] T. Piran and R. M. Williams, Inflation in universes with a massive scalar field, *Phys. Lett.* **B163**, pp. 331–335 (1985).

[8] T. Piran, On general conditions for inflation, *Phys. Lett.* **B181**, pp. 238–243 (1986), doi:10.1016/0370-2693(86)90039-0.

[9] J. J. Halliwell, Scalar fields in cosmology with an exponential potential, *Phys. Lett.* **B185**, p. 341 (1987), doi:10.1016/0370-2693(87)91011-2.

[10] F. Lucchin and S. Matarrese, Power law inflation, *Phys. Rev.* **D32**, p. 1316 (1985), doi:10.1103/PhysRevD.32.1316.

[11] J. D. Barrow, Cosmic no hair theorems and inflation, *Phys. Lett.* **B187**, pp. 12–16 (1987), doi:10.1016/0370-2693(87)90063-3.

[12] G. Ivanov, Friedmann's cosmological models with non-lenear scalar field, *Gravitaciya i Teoriya Otnositel'nosti* **18**, 1, pp. 54–60 (1981).

[13] K. A. Olive, Inflation, *Phys. Rept.* **190**, pp. 307–403 (1990a), doi:10.1016/0370-1573(90)90144-Q.

[14] K. A. Olive, Lectures on particle physics and cosmology, in *Proceedings, Summer School in High-energy Physics and Cosmology: Trieste, Italy, June 18-July 28, 1990*, pp. 421–494 (1990b).

[15] D. S. Goldwirth and T. Piran, Initial conditions for inflation, *Phys. Rept.* **214**, pp. 223–291 (1992), doi:10.1016/0370-1573(92)90073-9.

[16] A. Muslimov, On the scalar field dynamics in a spatially flat friedman universe, *Class. Quant. Grav.* **7**, pp. 231–237 (1990), doi:10.1088/0264-9381/7/2/015.

[17] J. D. Barrow, Graduated inflationary universes, *Phys. Lett.* **B235**, pp. 40–43 (1990), doi:10.1016/0370-2693(90)90093-L.

[18] G. F. R. Ellis and M. S. Madsen, Exact scalar field cosmologies, *Class. Quant. Grav.* **8**, pp. 667–676 (1991), doi:10.1088/0264-9381/8/4/012.

[19] S. V. Chervon and V. M. Zhuravlev, Exact solutions in cosmological inflationary models, *Russ. Phys. J.* **39**, pp. 776–780 (1996), doi:10.1007/BF02437088, [Izv. Vuz. Fiz.39N8,83(1996)].

[20] S. V. Chervon, V. M. Zhuravlev and V. K. Shchigolev, New exact solutions in standard inflationary models, *Phys. Lett.* **B398**, pp. 269–273 (1997), doi:10.1016/S0370-2693(97)00238-4, arXiv:gr-qc/9706031 [gr-qc].

[21] D. S. Salopek and J. R. Bond, Nonlinear evolution of long wavelength metric fluctuations in inflationary models, *Phys. Rev.* **D42**, pp. 3936–3962 (1990), doi:10.1103/PhysRevD.42.3936.

[22] W.-F. Wang, Exact solution in chaotic inflation model with potential minima, *Commun. Theor. Phys.* **36**, 1, pp. 122–124 (2001).

[23] D. Mitrinovitch, Analyse mathematique. Sur une equation differentielle du premier ordre intervenant dans divers problemes, *Comptes Rendus* **204**, pp. 1706–1708 (1937).

[24] V. M. Zhuravlev, S. V. Chervon and V. K. Shchigolev, New classes of exact solutions in inflationary cosmology, *J. Exp. Theor. Phys.* **87**, pp. 223–228 (1998), doi:10.1134/1.558649, [Zh. Eksp. Teor. Fiz.114,406(1998)].

[25] S. V. Chervon, V. K. Shchigolev and V. M. Zhuravlev, Nonlinear fields in models of cosmological inflation, *Russ. Phys. J.* **39**, pp. 139–145 (1996), doi:10.1007/BF02067677, [Izv. Vuz. Fiz.39N2,41(1996)].

[26] J. Barrow, Exact inflationary universes with potential minima, *Phys. Rev.* **D49**, pp. 3055–3058 (1994).

[27] J. D. Barrow and P. Parsons, Inflationary models with logarithmic potentials, *Phys. Rev.* **D52**, pp. 5576–5587 (1995), doi:10.1103/PhysRevD.52.5576, arXiv:astro-ph/9506049 [astro-ph].

[28] P. Parsons and J. D. Barrow, Generalized scalar field potentials and inflation, *Phys. Rev.* **D51**, pp. 6757–6763 (1995), doi:10.1103/PhysRevD.51.6757, arXiv:astro-ph/9501086 [astro-ph].

[29] V. F. Mukhanov, H. A. Feldman and R. H. Brandenberger, Theory of cosmological perturbations. Part 1. Classical perturbations. Part 2. Quantum theory of perturbations. Part 3. Extensions, *Phys. Rept.* **215**, pp. 203–333 (1992), doi:10.1016/0370-1573(92)90044-Z.

[30] J. D. Barrow, New types of inflationary universe, *Phys. Rev.* **D48**, pp. 1585–1590 (1993), doi:10.1103/PhysRevD.48.1585.

[31] E. Kamke, *Differentialgleichungen: Losungsmethoden und Losungen.* Chelsea Publishing Co, New York (1959).

[32] A. Chaadaev and S. V. Chervon, New class of cosmological solutions for a selfinteracting scalar field, *Russ. Phys. J.* **59**, pp. 725–730 (2013).

[33] S. Starkovich and F. Cooperstock, A cosmological field theory, *Astrophys. J.* **398**, pp. 1–11 (1992).

[34] A. R. Liddle, P. Parsons and J. D. Barrow, Formalizing the slow roll approximation in inflation, *Phys. Rev.* **D50**, pp. 7222–7232 (1994), doi:10.1103/PhysRevD.50.7222, arXiv:astro-ph/9408015 [astro-ph].

[35] A. H. Guth and B. Jain, Density fluctuations in extended inflation, *Phys. Rev.* **D45**, pp. 426–432 (1992), doi:10.1103/PhysRevD.45.426.

[36] A. R. Liddle and D. H. Lyth, COBE, gravitational waves, inflation and extended inflation, *Phys. Lett.* **B291**, pp. 391–398 (1992), doi:10.1016/0370-2693(92)91393-N, arXiv:astro-ph/9208007 [astro-ph].

[37] P. J. Steinhardt and M. S. Turner, A prescription for successful new inflation, *Phys. Rev.* **D29**, pp. 2162–2171 (1984), doi:10.1103/PhysRevD.29.2162.

[38] E. D. Stewart and D. H. Lyth, A more accurate analytic calculation of the spectrum of cosmological perturbations produced during inflation, *Phys. Lett.* **B302**, pp. 171–175 (1993), doi:10.1016/0370-2693(93)90379-V, arXiv:gr-qc/9302019 [gr-qc].

[39] E. J. Copeland, E. W. Kolb, A. R. Liddle and J. E. Lidsey, Reconstructing the inflation potential, in principle and in practice, *Phys. Rev.* **D48**, pp. 2529–2547 (1993), doi:10.1103/PhysRevD.48.2529, arXiv:hep-ph/9303288 [hep-ph].

[40] A. T. Kruger and J. W. Norbury, Another exact inflationary solution, *Phys. Rev.* **D61**, p. 087303 (2000), doi:10.1103/PhysRevD.61.087303, arXiv:gr-qc/0004039 [gr-qc].

[41] T. Harko, F. S. N. Lobo and M. K. Mak, Arbitrary scalar field and quintessence cosmological models, *Eur. Phys. J.* **C74**, p. 2784 (2014), doi:10.1140/epjc/s10052-014-2784-8, arXiv:1310.7167 [gr-qc].

[42] F. E. Schunck and E. W. Mielke, A New method of generating exact inflationary solutions, *Phys. Rev.* **D50**, pp. 4794–4806 (1994), doi:10.1103/PhysRevD.50.4794, arXiv:gr-qc/9407041 [gr-qc].

[43] H.-C. Kim, Exact solutions in Einstein cosmology with a scalar field, *Mod. Phys. Lett.* **A28**, p. 1350089 (2013), doi:10.1142/S0217732313500892.,10.1142/S0217732313500892, arXiv:1211.0604 [gr-qc].

[44] L. P. Chimento, A. E. Cossarini and A. S. Jakubi, Exact self-interacting scalar field cosmologies, in *Plebanski 65th Birthday Mexico City, Mexico, June 2-5, 1993*, pp. 2–4 (2012), arXiv:1208.0941 [gr-qc], http://inspirehep.net/record/1125950/files/arXiv:1208.0941.pdf.

[45] B. A. Bassett, S. Tsujikawa and D. Wands, Inflation dynamics and reheating, *Rev. Mod. Phys.* **78**, pp. 537–589 (2006), doi:10.1103/RevModPhys.78.537, arXiv:astro-ph/0507632 [astro-ph].

[46] N. Barbosa-Cendejas and M. A. Reyes, The Schrodinger picture of standard cosmology, (2010), arXiv:1001.0084 [gr-qc].

[47] J. A. S. Lima, J. A. M. Moreira and J. Santos, Particle like description for FRW cosmologies, *Gen. Rel. Grav.* **30**, pp. 425–434 (1998), doi:10.1023/A: 1018858809324.

[48] J. A. S. Lima, Note on solving for the dynamics of the universe, *Am. J. Phys.* **69**, pp. 1245–1247 (2001), doi:10.1119/1.1405506, arXiv:astro-ph/0109215 [astro-ph].

[49] Y. Leyva, D. Gonzalez, T. Gonzalez, T. Matos and I. Quiros, Dynamics of a self-interacting scalar field trapped in the braneworld for a wide variety of self-interaction potentials, *Phys. Rev.* **D80**, p. 044026 (2009), doi:10.1103/ PhysRevD.80.044026, arXiv:0909.0281 [gr-qc].

[50] R. M. Hawkins and J. E. Lidsey, The Ermakov-Pinney equation in scalar field cosmologies, *Phys. Rev.* **D66**, p. 023523 (2002), doi:10.1103/ PhysRevD.66.023523, arXiv:astro-ph/0112139 [astro-ph].

[51] F. L. Williams, P. G. Kevrekidis, T. Christodoulakis, C. Helias, G. O. Papadopoulos and T. Grammenos, On (3+1) dimensional scalar field cosmologies, (2004), arXiv:gr-qc/0408056 [gr-qc].

[52] T. Harko, F. S. N. Lobo and M. K. Mak, A Riccati equation based approach to isotropic scalar field cosmologies, *Int. J. Mod. Phys.* **D23**, 7, p. 1450063 (2014), doi:10.1142/S0218271814500631, arXiv:1402.4363 [gr-qc].

[53] S. V. Chervon and V. M. Zhuravlev, Comparative analysis of approximate and exact models in inflationary cosmology, *Russ. Phys. J.* **43**, pp. 11–17 (2000), doi:10.1007/BF02513001, [Izv. Vuz. Fiz.43N1,14(2000)].

[54] V. M. Zhuravlev and S. V. Chervon, Cosmological inflation models admitting natural emergence to the radiation-dominated stage and the matter domination Era, *J. Exp. Theor. Phys.* **91**, 2, pp. 227–238 (2000), doi: 10.1134/1.1311981, [Zh. Eksp. Teor. Fiz.118,no.2,259(2000)].

[55] E. Brezin, J. C. Le Guillou and J. Zinn-Justin, Perturbation theory at large order. 1. The ϕ^{2N} interaction, *Phys. Rev.* **D15**, pp. 1544–1557 (1977), doi: 10.1103/PhysRevD.15.1544.

[56] E. J. Copeland, A. R. Liddle and D. Wands, Exponential potentials and cosmological scaling solutions, *Phys. Rev.* **D57**, pp. 4686–4690 (1998), doi: 10.1103/PhysRevD.57.4686, arXiv:gr-qc/9711068 [gr-qc].

[57] T. Matos, J.-R. Luevano, I. Quiros, L. A. Urena-Lopez and J. A. Vazquez, Dynamics of scalar field dark matter with a cosh-like potential, *Phys. Rev.* **D80**, p. 123521 (2009), doi:10.1103/PhysRevD.80.123521, arXiv:0906.0396 [astro-ph.CO].

[58] M. A. Reyes, On exact solutions to the scalar field equations in standard cosmology, (2008), arXiv:0806.2292 [gr-qc].

[59] I. V. Fomin and S. V. Chervon, Exact and approximate solutions in the Friedmann cosmology, *Russ. Phys. J.* **60**, 3, pp. 427–440 (2017), doi:10.1007/s11182-017-1091-x.

[60] E. R. Harrison, Fluctuations at the threshold of classical cosmology, *Phys. Rev.* **D1**, pp. 2726–2730 (1970), doi:10.1103/PhysRevD.1.2726.

[61] Y. B. Zeldovich, Gravitational instability: An approximate theory for large density perturbations, *Astron. Astrophys.* **5**, pp. 84–89 (1970).

[62] J. M. Bardeen, Gauge invariant cosmological perturbations, *Phys. Rev.* **D22**, pp. 1882–1905 (1980), doi:10.1103/PhysRevD.22.1882.

[63] A. R. Liddle and D. H. Lyth, The cold dark matter density perturbation, *Phys. Rept.* **231**, pp. 1–105 (1993), doi:10.1016/0370-1573(93)90114-S, arXiv:astro-ph/9303019 [astro-ph].

[64] N. Straumann, From primordial quantum fluctuations to the anisotropies of the cosmic microwave background radiation, *Annalen Phys.* **15**, pp. 701–847 (2006), doi:10.1002/andp.200610212, arXiv:hep-ph/0505249 [hep-ph].

[65] V. N. Lukash, On the relation between tensor and scalar perturbation modes in Friedmann cosmology, *Usp. Fiz. Nauk* **176**, p. 113 (2006), doi: 10.1070/PU2006v049n01ABEH005901, arXiv:astro-ph/0610312 [astro-ph], [Phys. Usp.49,103(2006)].

[66] J. Garcia-Bellido, Cosmology and astrophysics, in *2004 European School of High-Energy Physics, Sant Feliu de Guixols, Spain, 30 May – 12 June 2004*, pp. 267–342 (2005), arXiv:astro-ph/0502139 [astro-ph], http://doc.cern.ch/yellowrep/2006/2006-003/p267.pdf.

[67] P. A. R. Ade *et al.*, Planck 2015 results. XIII. Cosmological parameters, *Astron. Astrophys.* **594**, p. A13 (2016), doi:10.1051/0004-6361/201525830, arXiv:1502.01589 [astro-ph.CO].

[68] D. Wands, K. A. Malik, D. H. Lyth and A. R. Liddle, A new approach to the evolution of cosmological perturbations on large scales, *Phys. Rev.* **D62**, p. 043527 (2000), doi:10.1103/PhysRevD.62.043527, arXiv:astro-ph/0003278 [astro-ph].

[69] J. A. Holtzman, Microwave background anisotropies and large scale structure in universes with cold dark matter, baryons, radiation, and massive and massless neutrinos, *Astrophys. J. Suppl.* **71**, pp. 1–24 (1989), doi: 10.1086/191362.

[70] S. Weinberg, Cosmological fluctuations of short wavelength, *Astrophys. J.* **581**, pp. 810–816 (2002), doi:10.1086/344441, arXiv:astro-ph/0207375 [astro-ph].

[71] A. V. Yurov, V. A. Yurov, S. V. Chervon and M. Sami, Total energy potential as a superpotential in integrable cosmological models, *Theor. Math. Phys.* **166**, pp. 259–269 (2011), doi:10.1007/s11232-011-0020-3.

[72] R. R. Caldwell, A Phantom menace? Cosmological consequences of a dark energy component with super-negative equation of state, *Phys. Lett.* **B545**, pp. 23–29 (2002), doi:10.1016/S0370-2693(02)02589-3, arXiv:astro-ph/9908168 [astro-ph].

[73] S. M. Carroll, M. Hoffman and M. Trodden, Can the dark energy equation-of-state parameter w be less than -1? *Phys. Rev.* **D68**, p. 023509 (2003), doi:10.1103/PhysRevD.68.023509, arXiv:astro-ph/0301273 [astro-ph].

[74] A. Yurov, Phantom scalar fields result in inflation rather than big rip, *Eur. Phys. J. Plus* **126**, p. 132 (2011), doi:10.1140/epjp/i2011-11132-7, arXiv:astro-ph/0305019 [astro-ph].

[75] N. Abel, *Oeuvres Complétes II*. S.Lie and L.Sylow, Eds., Christiana (1881).

[76] A. V. Yurov and V. A. Yurov, Friedman versus Abel equations: A connection unraveled, *J. Math. Phys.* **51**, p. 082503 (2010), doi:10.1063/1.3460856, arXiv:0809.1216 [hep-th].

[77] P. Appell, Sur les invariants de quelques équations différentielles, *Journal de Math.* **(4) 5**, pp. 361–423 (1889).

[78] R. Liouville, Sur une équation différentielle du premier ordre, *Acta Math.* **(4) 5**, pp. 55–78 (1903).

[79] G. Murphy, *Ordinary Differential Equations and Their Solution*. Van Nostrand, Princeton, NJ (1960).

[80] D. Zwillinger, *Handbook of Differential Equations, 3rd ed.* MA: Academic Press, Boston (1997).

[81] P. Sachdev, *A Compendium of Nonlinear Ordinary Differential Equations*. John Wiley & Sons, Michigan (1997).

[82] R. Easther, The evolution of scalar fields and inflationary cosmology, *Doctoral Dissertation, University of Canterbury* (1993).

[83] W. de Sitter, Einstein's theory of gravitation and its astronomical consequences (3rd paper), *Monthly Notices of the Royal Astronomical Society* **78**, pp. 3–28 (1917).

[84] A. D. Linde, Eternal Chaotic Inflation, *Mod. Phys. Lett.* **A1**, p. 81 (1986).

[85] A. A. Starobinsky, Spectrum of relict gravitational radiation and the early state of the universe, *JETP Lett.* **30**, pp. 682–685 (1979).

[86] J. Barrow and A. Ottewill, The stability of general relativistic cosmological theory, *J. Phys. A: Math. Gen.* **16**, pp. 2757–2776 (1983).

[87] S. Hawking, I. Moss and J. Stewart, Bubble collisions in the very early universe, *Phys. Rev.* **D26**, pp. 2681–2693 (1982).

[88] A. Guth and E. Weinberg, Could the universe have recovered from a slow first-order phase transition? *Nucl. Phys.* **B212**, pp. 321–364 (1983).

[89] A. Linde, *Particle Physics and Inflationary Cosmology*. Harwood Academic Publishers, Chur, Switzerland (1990).

[90] R. Maartens, D. Taylor and N. Roussos, Exact inflationary cosmologies with exit, *Phys. Rev.* **D52**, pp. 3358–3364 (1995).

[91] A. Yurov, Exact inflationary cosmologies with exit: From an inflaton complex field to an "anti-inflaton" one, *Class. Quantum Grav.* **18**, pp. 3753–3766 (2001).

[92] A. V. Yurov and S. D. Vereshchagin, The Darboux transformation and exactly solvable cosmological models, *Theor. Math. Phys.* **139**, pp. 787–800 (2004), doi:10.1023/B:TAMP.0000029701.48809.a1, arXiv:hep-th/0502099 [hep-th], [Teor. Mat. Fiz.139,405(2004)].

[93] S. Hannestad and E. Mortsell, Probing the dark side: Constraints on the dark energy equation of state from CMB, large scale structure and Type Ia supernovae, *Phys. Rev.* **D66**, p. 063508 (2002), doi:10.1103/PhysRevD.66.063508, arXiv:astro-ph/0205096 [astro-ph].

[94] A. C. Doyle, *The Sign of the Four*. Spencer Blackett (1890).

[95] P. F. Gonzalez-Diaz, Achronal cosmic future, *Phys. Rev. Lett.* **93**, p. 071301 (2004), doi:10.1103/PhysRevLett.93.071301, arXiv:astro-ph/0404045 [astro-ph].

[96] S. Perlmutter *et al.*, Discovery of a supernova explosion at half the age of the universe and its cosmological implications, *Nature* **391**, pp. 51–54 (1998), doi:10.1038/34124, arXiv:astro-ph/9712212 [astro-ph].

[97] B. McInnes, The dS/CFT correspondence and the big smash, *JHEP* **08**, p. 029 (2002), doi:10.1088/1126-6708/2002/08/029, arXiv:hep-th/0112066 [hep-th].

[98] P. F. Gonzalez-Diaz and J. A. Jimenez-Madrid, Phantom inflation and the 'big trip', *Phys. Lett.* **B596**, pp. 16–25 (2004), doi:10.1016/j.physletb.2004.06.080, arXiv:hep-th/0406261 [hep-th].

[99] F. S. N. Lobo, Phantom energy traversable wormholes, *Phys. Rev.* **D71**, p. 084011 (2005), doi:10.1103/PhysRevD.71.084011, arXiv:gr-qc/0502099 [gr-qc].

[100] P. Martin-Moruno, J. A. J. Madrid and P. F. Gonzalez-Diaz, Will black holes eventually engulf the universe? *Phys. Lett.* **B640**, pp. 117–120 (2006), doi:10.1016/j.physletb.2006.07.067, arXiv:astro-ph/0603761 [astro-ph].

[101] L. Randall and R. Sundrum, An alternative to compactification, *Phys. Rev. Lett.* **83**, pp. 4690–4693 (1999), doi:10.1103/PhysRevLett.83.4690, arXiv:hep-th/9906064 [hep-th].

[102] S. V. Krasnikov, Superluminal motion in (semi)classical relativity, (2006), arXiv:gr-qc/0603060 [gr-qc].

[103] S. V. Krasnikov, Hyperfast travel in general relativity, *Phys. Rev.* **D57**, pp. 4760–4766 (1998), doi:10.1103/PhysRevD.57.4760, arXiv:gr-qc/9511068 [gr-qc].

[104] A. E. Everett and T. A. Roman, A superluminal subway: The Krasnikov tube, *Phys. Rev.* **D56**, pp. 2100–2108 (1997), doi:10.1103/PhysRevD.56.2100, arXiv:gr-qc/9702049 [gr-qc].

[105] L. H. Ford and T. A. Roman, Quantum field theory constrains traversable wormhole geometries, *Phys. Rev.* **D53**, pp. 5496–5507 (1996), doi:10.1103/PhysRevD.53.5496, arXiv:gr-qc/9510071 [gr-qc].

[106] L. H. Ford, M. J. Pfenning and T. A. Roman, Quantum inequalities and singular negative energy densities, *Phys. Rev.* **D57**, pp. 4839–4846 (1998), doi:10.1103/PhysRevD.57.4839, arXiv:gr-qc/9711030 [gr-qc].

[107] D. N. Vollick, Quantum inequalities in curved two-dimensional space-times, *Phys. Rev.* **D61**, p. 084022 (2000), doi:10.1103/PhysRevD.61.084022, arXiv:gr-qc/0001009 [gr-qc].

[108] C. J. Fewster, Quantum energy inequalities in two dimensions, *Phys. Rev.* **D70**, p. 127501 (2004), doi:10.1103/PhysRevD.70.127501, arXiv:gr-qc/0411114 [gr-qc].

[109] M. J. Pfenning and L. H. Ford, The unphysical nature of 'warp drive', *Class. Quant. Grav.* **14**, pp. 1743–1751 (1997), doi:10.1088/0264-9381/14/7/011, arXiv:gr-qc/9702026 [gr-qc].

[110] K. Godel, An example of a new type of cosmological solutions of Einstein's field equations of gravitation, *Rev. Mod. Phys.* **21**, pp. 447–450 (1949).

[111] E. Babichev, V. Dokuchaev and Yu. Eroshenko, Black hole mass decreasing due to phantom energy accretion, *Phys. Rev. Lett.* **93**, p. 021102 (2004), doi:10.1103/PhysRevLett.93.021102, arXiv:gr-qc/0402089 [gr-qc].

[112] R. Maartens, Brane world gravity, *Living Rev. Rel.* **7**, p. 7 (2004), arXiv:gr-qc/0312059 [gr-qc].

[113] M. S. Morris, K. S. Thorne and U. Yurtsever, Wormholes, time machines, and the weak energy condition, *Phys. Rev. Lett.* **61**, pp. 1446–1449 (1988), doi:10.1103/PhysRevLett.61.1446.

[114] S. W. Hawking, The chronology protection conjecture, *Phys. Rev.* **D46**, pp. 603–611 (1992), doi:10.1103/PhysRevD.46.603.

[115] S. Krasnikov, The time travel paradox, *Phys. Rev.* **D65**, p. 064013 (2002), doi:10.1103/PhysRevD.65.064013, arXiv:gr-qc/0109029 [gr-qc].

[116] D. Deutsch, Quantum mechanics near closed timelike lines, *Phys. Rev.* **D44**, pp. 3197–3217 (1991), doi:10.1103/PhysRevD.44.3197.

[117] A. Everett, Time travel paradoxes, path integrals, and the many worlds interpretation of quantum mechanics, *Phys. Rev.* **D69**, p. 124023 (2004), doi:10.1103/PhysRevD.69.124023, arXiv:gr-qc/0410035 [gr-qc].

[118] A. Carlini, V. Frolov, M. Mensky, I. Novikov and H. Soleng, Time machines and the principle of self-consistency as a consequence of the Principle of Stationary Action (II): The Cauchy problem for a self-interacting relativistic particle, *Int. J. Mod. Phys.* **D5**, pp. 445–480 (1996).

[119] D. Bohm, A Suggested interpretation of the quantum theory in terms of hidden variables. 2. *Phys. Rev.* **85**, pp. 180–193 (1952), doi:10.1103/PhysRev.85.180.

[120] F. J. Tipler, Genesis: How the universe began according to standard model particle physics, (2001), arXiv:astro-ph/0111520 [astro-ph].

[121] J. B. Hartle and S. W. Hawking, Wave function of the universe, *Phys. Rev.* **D28**, pp. 2960–2975 (1983), doi:10.1103/PhysRevD.28.2960.

[122] J. R. Gott, III and L.-X. Li, Can the universe create itself? *Phys. Rev.* **D58**, p. 023501 (1998), doi:10.1103/PhysRevD.58.023501, arXiv:astro-ph/9712344 [astro-ph].

[123] P. F. Gonzalez-Diaz, Holographic cosmic energy, fundamental theories and the future of the universe, *Grav. Cosmol.* **12**, pp. 29–36 (2006), arXiv:astro-ph/0507714 [astro-ph].

[124] F. Wilczek, Analysis and synthesis III: Cosmic groundwork, *Phys. Today* **56N10** (2003).

[125] D. Goldsmith, Einstein's greatest blunder, *Harvard University Press* (1997).

[126] J. Halliwell, J. Perez-Mercader and W. Zurek, The physical origin of time asymmetry, *Cambridge University Press* (1994).

[127] T. Banks, Cosmological breaking of supersymmetry? *Int. J. Mod. Phys.* **A16**, pp. 910–921 (2001), doi:10.1142/S0217751X01003998, arXiv:hep-th/0007146 [hep-th], [,270(2000)].

[128] H. E. III, J. Wheeler and B. DeWitt, *The Many-Worlds Interpretation of Quantum Mechanics*, Princeton University Press (1973).

[129] L. Susskind, The Anthropic Landscape of String Theory, (2003), arXiv:hep-th/0302219 [hep-th].

[130] L. Smolin, *The Life of the Cosmos*. Oxford University Press (1997).

[131] S. Robles-Perez, P. Martin-Moruno, A. Rozas-Fernandez and P. Gonzalez-Daz, A dark energy multiverse, *Quant. Grav.* **24**, p. 3 (2007), doi:10.1088/0264-9381/24/10/F01.

[132] A. Bloom, *The Republic of Plato*. Basic Books, New York, USA (1991).

[133] D. N. Spergel *et al.*, Wilkinson Microwave Anisotropy Probe (WMAP) three year results: Implications for cosmology, *Astrophys. J. Suppl.* **170**, p. 377 (2007), doi:10.1086/513700, arXiv:astro-ph/0603449 [astro-ph].

[134] S. Hawking and G. Ellis, *The Large Scalestructure of Space-time*. Cambridge University Press, Cambridge, UK (1973).

[135] W. L. Freedman *et al.*, Final results from the Hubble Space Telescope key project to measure the Hubble constant, *Astrophys. J.* **553**, pp. 47–72 (2001), doi:10.1086/320638, arXiv:astro-ph/0012376 [astro-ph].

[136] A. Yurov, P. Martin-Moruno and P. Gonzalez-Diaz, New bigs in cosmology, *Phys.* **B759**, p. 320341 (2006), doi:10.1088/0264-9381/24/10/F01.

[137] L. Randall and R. Sundrum, A Large mass hierarchy from a small extra dimension, *Phys. Rev. Lett.* **83**, pp. 3370–3373 (1999), doi:10.1103/PhysRevLett.83.3370, arXiv:hep-ph/9905221 [hep-ph].

[138] E. E. Flanagan, S. H. H. Tye and I. Wasserman, Brane world models with bulk scalar fields, *Phys. Lett.* **B522**, pp. 155–165 (2001), doi:10.1016/S0370-2693(01)01261-8, arXiv:hep-th/0110070 [hep-th].

[139] S. Perlmutter *et al.*, Measurements of omega and lambda from 42 high redshift supernovae, *Astrophys. J.* **517**, pp. 565–586 (1999), doi:10.1086/307221, arXiv:astro-ph/9812133 [astro-ph].

[140] P. M. Garnavich *et al.*, Supernova limits on the cosmic equation of state, *Astrophys. J.* **509**, pp. 74–79 (1998), doi:10.1086/306495, arXiv:astro-ph/9806396 [astro-ph].

[141] S. H. H. Tye and I. Wasserman, A Brane world solution to the cosmological constant problem, *Phys. Rev. Lett.* **86**, pp. 1682–1685 (2001), doi:10.1103/PhysRevLett.86.1682, arXiv:hep-th/0006068 [hep-th].

[142] O. DeWolfe, D. Z. Freedman, S. S. Gubser and A. Karch, Modeling the fifth-dimension with scalars and gravity, *Phys. Rev.* **D62**, p. 046008 (2000), doi:10.1103/PhysRevD.62.046008, arXiv:hep-th/9909134 [hep-th].

[143] Z. Kakushadze, Localized (super)gravity and cosmological constant, *Nucl. Phys.* **B589**, pp. 75–118 (2000), doi:10.1016/S0550-3213(00)00514-9, arXiv:hep-th/0005217 [hep-th].

[144] V. Matveev and M. Salle, *Darboux Transformation and Solitons*. Springer Verlag, Berlin–Heidelberg (1991).

[145] A. A. Andrianov, N. V. Borisov and M. V. Ioffe, The factorization method and quantum systems with equivalent energy spectra, *Phys. Lett.* **A105**, pp. 19–22 (1984), doi:10.1016/0375-9601(84)90553-X.

[146] A. A. Andrianov, N. V. Borisov, M. I. Eides and M. V. Ioffe, Supersymmetric origin of equivalent quantum systems, *Phys. Lett.* **A109**, pp. 143–148 (1985), doi:10.1016/0375-9601(85)90004-0.

[147] V. E. Adler, A modification of Crum's method, *Theor. and Math. Phys.* **101**, pp. 1381–1386 (1994).

[148] J. G. Darboux, Sur une proposition relative aux equations lineaires, *Compt. Rend.* **94**, p. 1343 (1882).

[149] B. P. Berezovoy and A. I. Pashnev, Supersymmetric quantum mechanics and rearrangement of the spectra of Hamiltonians, *Theor. and Math. Phys.* **70**, pp. 102–107 (1987).

[150] W. Israel, Singular hypersurfaces and thin shells in general relativity, *Nuovo Cimento* **44 (S10)**, pp. 1–14 (1966).

[151] V. A. Rubakov, Large and infnite extra dimensions: An introduction, *Uspehi Fiz. Nauk* **171**, p. 913 (2001).

[152] M. M. Crum, Associated Sturm-Liouville Systems, *Quart. J. Math. Ser. Oxford* **6**, pp. 121–127 (1955).

[153] L. Infeld and T. E. Hull, The factorization method, *Rev. Mod. Phys.* **23**, pp. 21–68 (1951), doi:10.1103/RevModPhys.23.21.

[154] A. Veselov and A. Shabat, Dressing chains and the spectral theory of the schrodinger operator, *Funct. Anal. Appl.* **27(2)**, pp. 81–96 (1993).

[155] V. Bargmann, On the connection between phase shifts and scattering potential, *Rev. Mod. Phys.* **21**, pp. 488–493 (1949).

[156] A. Gonzalez-Lopez and N. Kamran, The multidimensional Darboux transformation, *J. Geom. Phys.* **26**, pp. 202–226 (1998), doi:10.1016/S0393-0440(97)00044-2, arXiv:hep-th/9612100 [hep-th].

[157] A. G. Riess *et al.*, Observational evidence from supernovae for an accelerating universe and a cosmological constant, *Astron. J.* **116**, pp. 1009–1038 (1998), doi:10.1086/300499, arXiv:astro-ph/9805201 [astro-ph].

[158] A. Starobinsky, Future and origin of our universe: Modern view, *Grav. Cosmol* **6**, pp. 157–163 (2000), arXiv:astro-ph/9912054 [astro-ph].

[159] M. Bouhmadi-Lopez, P. F. Gonzalez-Diaz and P. Martin-Moruno, Worse than a big rip? *Phys. Rev.* **B659**, pp. 1–5 (2008), doi:10.1016/j.physletb.2007.10.079, arXiv:gr-qc/0612135 [gr-qc].

[160] Y. Shtanov and V. Sahni, New cosmological singularities in braneworld models? *Class. Quant. Grav.* **19**, pp. L101–L107 (2002), doi:10.1088/0264-9381/19/11/102, arXiv:gr-qc/0204040 [gr-qc].

[161] J. D. Barrow, Sudden future singularities, *Class. Quant. Grav.* **21**, pp. L79–L82 (2004), doi:10.1088/0264-9381/21/11/L03, arXiv:gr-qc/0403084 [gr-qc].

[162] A. Barvinsky, C. Deffayet and A. Kamenshchik, Anomaly driven cosmology: Big boost scenario and AdS/CFT correspondence, *JCAP* **0805**, p. 020 (2008), doi:10.1088/1475-7516/2008/05/020, arXiv:0801.2063 [hep-th].

[163] V. Gorini, A. Y. Kamenshchik, U. Moschella and V. Pasquier, Tachyons, scalar fields and cosmology, *Phys. Rev.* **D69**, p. 123512 (2004), doi:10.1103/PhysRevD.69.123512, arXiv:hep-th/0311111 [hep-th].

[164] A. Kamenshchik, C. Kiefer and B. Sandhoefer, Quantum cosmology with big-brake singularity, *Phys. Rev.* **D76**, p. 064032 (2007), doi:10.1103/ PhysRevD.76.064032, arXiv:0705.1688 [gr-qc].

[165] S. Nojiri and S. D. Odintsov, Singularity of spherically-symmetric spacetime in quintessence/phantom dark energy universe, *Phys. Lett.* **B676**, pp. 94–98 (2009), doi:10.1016/j.physletb.2009.04.079, arXiv:0903.5231 [hep-th].

[166] S. Nojiri and S. D. Odintsov, The future evolution and finite-time singularities in $F(R)$-gravity unifying the inflation and cosmic acceleration, *Phys. Rev.* **D78**, p. 046006 (2008), doi:10.1103/PhysRevD.78.046006, arXiv:0804.3519 [hep-th].

[167] S. Capozziello, M. D. Laurentis, S. Nojiri and S. Odintsov, Classifying and avoiding singularities in the alternative gravity dark energy models, *Phys. Rev.* **D79**, p. 124007 (2009), doi:10.1103/PhysRevD.79.124007, arXiv:0903.2753 [hep-th].

[168] S. Nojiri, S. D. Odintsov and S. Tsujikawa, Classifying and avoiding singularities in the alternative gravity dark energy models, *Phys. Rev.* **D71**, p. 063004 (2005), doi:10.1103/PhysRevD.71.063004, arXiv:hep-th/0501025 [hep-th].

[169] M. P. Dabrowski and T. Denkiewicz, Barotropic index w-singularities in cosmology, *Phys. Rev.* **D 79**, p. 063521 (2009), doi:10.1103/PhysRevD.79. 063521, arXiv:0902.3107 [gr-qc].

[170] M. P. Dabrowski, T. Denkiewicz and M. A. Hendry, How far is it to a sudden future singularity of pressure? *Phys. Rev.* **D75**, p. 123524 (2007), doi:10.1103/PhysRevD.75.123524, arXiv:0704.1383 [astro-ph].

[171] S. Nojiri and S. D. Odintsov, Quantum escape of sudden future singularity, *Phys. Lett.* **B595**, pp. 1–8 (2004), doi:10.1016/j.physletb.2004.06.060, arXiv:hep-th/0405078 [hep-th].

[172] A. Yurov, A. Astashenov and V. Yurov, The dressing procedure for the cosmological equations and the indefinite future of the universe, *Grav. Cosmol.* **14**, pp. 8–16 (2008), doi:10.1134/S0202289308010027, arXiv: astro-ph/0701597 [astro-ph/].

[173] A. Yurov and A. Astashenok, The linearization method and new classes of exact solutions in cosmology, *Theor. Math. Phys.* **158**, pp. 261–268 (2009), doi:10.1007/s11232-009-0021-7, arXiv:0902.1979 [astro-ph].

[174] A. Sen, Rolling Tachyon, *JHEP* **0204**, p. 048 (2002), doi:10.1088/ 1126-6708/2002/04/048, arXiv:hep-th/0203211 [hep-th].

[175] P. Binetruy, C. Deffayet and D. Langlois, Nonconventional cosmology from a brane universe, *Nucl. Phys.* **B565**, pp. 269–287 (2000a), doi:10.1016/ S0550-3213(99)00696-3, arXiv:hep-th/9905012 [hep-th].

[176] P. Binetruy, C. Deffayet, U. Ellwanger and D. Langlois, Brane cosmological evolution in a bulk with cosmological constant, *Phys. Lett.* **B477(1)**, pp. 285–291 (2000b).

[177] S. Mukohyama, Brane-world solutions, standard cosmology, and dark radiation, *Phys. Lett.* **B473**, pp. 241–245 (2000), doi:10.1016/S0370-2693(99) 01505-1, arXiv:hep-th/9911165 [hep-th].

[178] D. N. Vollick, Cosmology on a three-brane, *Class. Quant. Grav.* **18**, pp. 1–10 (2001), doi:10.1088/0264-9381/18/1/301, arXiv:hep-th/9911181 [hep-th].

[179] P. Kraus, Dynamics of Anti-de Sitter domain walls, *JHEP* **9912**, p. 011 (1999), doi:10.1088/1126-6708/1999/12/011, arXiv:hep-th/9910149 [hep-th].

[180] D. Ida, Brane-world cosmology, *JHEP* **0009**, p. 014 (2000), doi:10.1088/1126-6708/2000/09/014, arXiv:gr-qc/9912002 [gr-qc].

[181] S. Mukohyama, T. Shiromizu and K. Maeda, Global structure of exact cosmological solutions in the brane world, *Phys. Rev.* **D62**, p. 024028 (2000), doi:10.1103/PhysRevD.63.029901, 10.1103/PhysRevD.62.024028, arXiv:hep-th/9912287 [hep-th], [Erratum: Phys. Rev.D63,029901(2001)].

[182] B. Paul and D. Paul, Brane World Inflation with Scalar and Tachyon Fields, (2007), arXiv:0708.0897 [hep-th].

[183] K. A. Bronnikov and B. E. Meierovich, A general thick brane supported by a scalar field, *Grav. Cosmol.* **9**, pp. 313–318 (2003), arXiv:gr-qc/0402030 [gr-qc].

[184] S. Mizuno, S.-J. Lee and E. J. Copeland, Cosmological evolution of general scalar fields in a brane-world cosmology, *Phys. Rev.* **D70**, p. 043525 (2004), doi:10.1103/PhysRevD.70.043525, arXiv:astro-ph/0405490 [astro-ph].

[185] S. Chervon and M. Sami, Exact solutions of cosmological inflation on the Randall-Sundrum brane, *Electronnyi Zhurnal "Issledovano v Rossii"* **088/091009**, p. 1151 (2009).

[186] V. Sahni and Y. Shtanov, Cosmic Acceleration and Extra Dimensions, (2008), arXiv:0811.3839 [astro-ph].

[187] D. Langlois, Brane cosmology: An introduction, *Prog. Teor. Phys. Suppl.* **148**, p. 181 (2003).

[188] E. Copeland, M. Sami and S. Tsujikawa, Dynamics of dark energy, *Int. J. Mod. Phys.* **D15**, pp. 1753–1936 (2006), doi:10.1142/S021827180600942X, arXiv:hep-th/0603057 [hep-th].

[189] E. J. Copeland, A. R. Liddle and J. E. Lidsey, Steep inflation: Ending braneworld inflation by gravitational particle production, *Phys. Rev.* **D64**, p. 023509 (2001), doi:10.1103/PhysRevD.64.023509, arXiv:astro-ph/0006421 [astro-ph].

[190] V. Sahni, M. Sami and T. Souradeep, Relic gravity waves from braneworld inflation, *Phys. Rev.* **D65**, p. 023518 (2002), doi:10.1103/PhysRevD.65.023518, arXiv:gr-qc/0105121 [gr-qc].

[191] M. Sami and V. Sahni, Quintessential inflation on the brane and the relic gravity wave background, *Phys. Rev.* **D70**, p. 083513 (2004), doi:10.1103/PhysRevD.70.083513, arXiv:hep-th/0402086 [hep-th].

[192] Y. Shtanov and V. Sahni, Bouncing braneworlds, *Phys. Lett.* **B557**, pp. 1–6 (2003), doi:10.1016/S0370-2693(03)00179-5, arXiv:gr-qc/0208047 [gr-qc].

[193] E. J. Copeland, S.-J. Lee, J. E. Lidsey and S. Mizuno, Generalised cosmological scaling solutions, *Phys. Rev.* **D71**, p. 023526 (2005), doi:10.1103/PhysRevD.71.023526, arXiv:astro-ph/0410110 [astro-ph].

[194] A. Ashtekar, T. Pawlowski and P. Singh, Quantum nature of the big bang: Improved dynamics, *Phys. Rev.* **D74**, p. 084003 (2006), doi:10.1103/PhysRevD.74.084003, arXiv:gr-qc/0607039 [gr-qc].

[195] N. Kanekar, V. Sahni and Y. Shtanov, Recycling the universe using scalar fields, *Phys. Rev.* **D63**, p. 083520 (2001), doi:10.1103/PhysRevD.63.083520, arXiv:astro-ph/0101448 [astro-ph].

[196] S. Chervon, Inflationary cosmological models without restrictions on a scalar field potential, *Gen. Rel. Grav.* **36**, pp. 1547–1553 (2004), doi:10.1023/B:GERG.0000032147.11600.d2.

[197] S. V. Chervon and I. V. Fomin, On calculation of the cosmological parameters in exact models of inflation, *Grav. Cosmol.* **14**, pp. 163–167 (2008), doi:10.1134/S0202289308020060.

Index

Series on the Foundations of Natural Science and Technology

(Continued from page ii)